Gene Delivery to the
Blood-Brain Barrier

Gene Delivery to the Blood-Brain Barrier

*Non-Viral Gene Carriers and in vitro
Blood-Brain Barrier Models*

PhD Thesis by

Louiza Bohn Thomsen

*Laboratory for Neurobiology – Biomedicine,
Department of Health Science and Technology,
Aalborg University, Aalborg, Denmark*

River Publishers

Aalborg

ISBN 978-87-92982-14-8 (paperback)
ISBN 978-87-92982-15-5 (e-book)

Published, sold and distributed by:
River Publishers
P.O. Box 1657
Algade 42
9000 Aalborg
Denmark

Tel.: +45369953197
www.riverpublishers.com

Table of contents

Preface

This thesis has been submitted to the Faculty of Medicine, Aalborg University, Denmark. The work in this thesis has been carried out in the Laboratory of Neurobiology, Biomedicine, Department of Health Science and Technology, Aalborg University from October 2008 until November 2011. In November 2009, I visited Dr. Damir Janigro and his research group on Cleveland Clinic, Ohio, USA and worked shortly in his laboratory to learn how to operate their dynamic *in vitro* blood-brain barrier system, Flocel. During my Ph.D. study I have participated in Ph.D. courses corresponding to half a year. Furthermore, I have been teaching students at the educations "Medicine with industrial specialization" and "Medicine" according to the norms of Aalborg University which also corresponds to half a year of studying.

This thesis contains the following: Introduction, Objective of the Thesis, Results, and Discussion. The results are presented in article form in three studies. Study I was published in Journal of Controlled Release in 2010. Study II has been submitted to Journal of Controlled Release and Study III are yet unpublished. The thesis also contains two review articles appendices I and II published or accepted for publication, respectively, during my Ph.D. study.

Aalborg, November 2011

Louiza Bohn Thomsen

Louiza Bohn Thomsen

Supervisors

Head Supervisor:
Professor Torben Moos
Department of Health Science and Technology
Aalborg University
Aalborg, Denmark

Co-Supervisor:
Professor Thomas Jensen
Department of Human Genetics
University of Aarhus
Aarhus, Denmark

Co-Supervisor:
Anker Jon Hansen
Principal Scientist at Novo Nordisk
Novo Nordisk A/S
Bagsværd, Denmark

Abbreviations

ACM: astrocyte-conditioned media
ANG1: angiopoitin 1
BBB: blood-brain barrier
BCECs: brain capillary endothelial cells
BDNF: brain derived neurotropic factor
bFGF: basic fibroblast growth factor
cDNA: complementary DNA
CED: convection enhanced diffusion
CMV: cytomegalovirus
CSF: cerebrospinal fluid
EPO: Erythropoietin
GDNF: Glial cell line-derived neurotropic factor
GH: growth hormone
hGH1: human growth hormone 1
L-DOPA: L-3,4 dihydroxyphenylalanine
MRI: magnetic resonance imaging
NGF: neural growth factor
NLS: nuclear localization signal
PECAM: platelet endothelial cell adhesion molecule
PEG: polyethylene glycol
PEI: polyethylenimine
siRNA: short interfering RNA
SPIO: superparamagnetic iron oxide nanoparticle
ZO-1: zonula occludens 1

Abstract

Drug- and gene delivery to the brain is highly restricted by the vascular barriers of the brain, denoted by the blood-brain barrier (BBB) and the blood-cerebrospinal fluid (CSF) barriers. Among these barriers, BBB is the main limiting factor. It is composed by the brain capillary endothelial cells (BCECs). The BCECs barrier function is supported by astrocytes, pericytes and neurons to form the blood-brain barrier. BCECs are very tightly connected to each other by tight junctions. Apart from the essential substrates needed to nourish the brain, small and/or lipophilic molecules are free to diffuse into the brain. However most pharmacologically active drugs and gene fragments are too large to enter the brain. Various kinds of drug-carriers have been constructed for delivery of large substances to the brain. Such drug-carriers have to be able to successfully carry their cargo through the BCECs and into the brain. For testing the ability of drug-carriers to deliver their cargo into the brain, investigators have constructed different *in vitro* BBB models, consisting of BCECs that express the main characteristics of the BBB *in vivo*.

In the first part of the thesis the ability of two drug-carriers, pullulan-spermine and SPIOs, to mediate transfection of BCECs or transcellular transport through BCECs *in vitro* was studied.

Pullulan-spermine is a polymeric complex consisting of the polysaccharide, pullulan and the polyamine, spermine. Pullulan-spermine formed a cationic complex shown to be able to bind plasmid DNA electrostatically. Pullulan-spermine was conjugated with plasmid DNA encoding a red fluorescent protein, Hc-Red-1 C1, or human growth hormone 1 (hGH1). Pullulan-spermine complexed with Hc-Red-1 C1 cDNA led to the formation of a red fluorescent signal in human brain microvascular endothelial cells (HBMECs). Furthermore, pullulan-spermine complexed with hGH1 cDNA was not only able to transfect HBMECs but also led to secretion of the hGH1 into the culture media. Pullulan-spermine-cDNA complexes could transfect non-dividing cells although the rate of transgene cells was higher in dividing cells. This indicated that the DNA is not only entering the cell nucleus under mitosis. Unfortunately, pullulan-spermine complexes proved incapable of transfecting HBMECs in the presence of serum in the growth media and additional studies are needed to enable its use for *in vivo* transfection.

Another potential drug-carrier, fluorescent iron oxide nanoparticles were also shown to enter HBMECs upon incubation. These nanoparticles were also able to pass though the HBMECs forming a BBB in a static *in vitro* BBB model. Furthermore, their passage was increased by the aid of an external magnetic field created by placing the cell culture plates with the SPIOs on a plate magnet. Two vitality tests showed no significant change in BCEC vitality after addition of

SPIOs or by dragging the nanoparticles through the BCECs in the presence of the external electric field.

The results of the drug-carrier studies indicate that it is possible to deliver plasmid cDNA into BCECs and transfect these cells leading to their secretion of encoded protein into the extracellular space. Moreover, SPIOs are potentially potent carriers of attachable molecules trough cultured BCECs *in vitro*, which may have high potential for drug-delivery to the brain *in vivo*.

In the second part of the thesis, two in vitro BBB models, a static and a dynamic model was investigated and compared. The static model consisting of microporous membrane inserts in which immortalized BCECs is cultured. The model induces many characteristics of the BBB *in vivo*, but lacks the tightness induction factor of shear stress. Different experiments were performed with this static model to monitor BBB integrity. Barrier formation by the BCECs was monitored by measuring transendothelial electric resistance (TEER) and the BCEC monolayer was stained positive for zonula occludens 1 (ZO-1) a tight junction protein. It was mainly found that the tightness of the BCECs was strengthened by contact co-culture of the BCECs with astrocytes and addition of hydrocortisone to the media. The dynamic *in vitro* BBB model however, did not lead to any reliable results in this study and further investigation of barrier formation in this model was not pursued. In consequence a comparison between the static and dynamic *in vitro* models was not possible, but it could be concluded that the static model seems to be the most reliable model.

Resumé på dansk

Gen og medikament levering til hjernen har vist sig at være svært hæmmet af barriererne i hjernen, herunder hovedsageligt blod-hjerne barrieren (BBB). BBB formes af hjerne kapillær endothel celler (BCECs). BCECs er omgivet af astrocytter, pericytter og neuroner, der også menes at støtte BCECs barriere funktioner. BCECs danner tætte forbindelser mellem hinanden, kaldet "tight-junctions" og derved forhindres passage mellem cellerne. Bortset fra de essentielle næringsstoffer hjernen har brug for er det kun små og/eller fedtopløselige molekyler, der kan passere BBB. Gener og medikamenter er ofte store og vandopløselige og kan derfor ikke passere gennem BBB. Derfor er der brug for en leverings strategi af store molekyler til hjernen. Forskellige "drug carriers" er blevet udviklet til dette formål. Drug carriers bør være i stand til at levere deres last gennem BCECs og videre ind i hjernen. Til at teste en drug carriers evne bruger man ofte en *in vitro* BBB model. Disse modeller består af BCECs dyrket i kultur, der danner en barriere med de karakteristika BBB udviser *in vivo*.

I den første del af denne tese blev to potentielle drug-carriers undersøgt for deres evner til enten at passere BBB og ind i hjernen eller at levere gener ind i BCECs.

Den første carrier er et polymerisk kompleks bestående af en polysakkarid, pullulan og en polyaminosyre, spermine. Pullulan-spermine danner tilsammen et kationisk kompleks der kan binde negativt ladet plasmid DNA elektrostatisk. I dette studie blev pullulan-spermine konjugeret med plasmid DNA kodende for en rød fluorescerende markør Hc-Red C1 og humant vækst hormon 1 (hGH1). Pullulan-spermine-pHc-Red-1 C1 komplekser havde evnen til at forme transgene rød fluorescerende humane hjerne endothel celler (HBMECs) i monokultur. Endvidere kunne pullulan-spermine-pGH1 komplekser transfectere HBMECs og proteinet som det plasmide DNA kodede for, hGH1, kunne detekteres i cellerne og i celle kultur mediet, hvilket indikerede at HBMECs havde udskilt dette protein. Pullulan-spermine viste sig desværre at være uforenelig med serum, hvilket forhindrer brug af denne carrier *in vivo*. Derudover blev det fundet at celler der ikke er i det delende stadie, dvs. ikke mitotiske, kunne transfecteres, dog i en mindre grad end i delende celler. Dette indikerer at det plasmide DNA ikke kun kan diffundere ind i cellekernen, når kernemembranen er midlertidigt åben, men at der også findes en mekanisme der kan hjælpe det plasmide DNA ind i cellekernen når kernemembranen er intakt. Resultaterne i dette studie indikerer at det er muligt at benytte pullulan-spermine komplekser til levering af DNA til BCECs *in vitro* og at disse kan transfecteres og udskille det DNA indkodede protein.

Den anden carrier type hvis evner blev undersøgt I dette studie var fluorescerende superparamagnetiske nanopartikler (SPIOs). I dette studie kunne det påvises at fluorescerende stivelses overflade behandlede jern oxid nanopartikler kunne optages af BCECs. Desuden kunne disse partikler passere BCECs, der dannede en BBB i en statisk *in vitro* BBB model. Derudover kunne det påvises at passage af SPIOs øges markant ved hjælp af et eksternt magnetisk felt. Dette eksterne magnetiske felt blev dannet ved hjælp af en plademagnet hvorpå cellekultur pladerne med BCECs blev placeret og de magnetiske nanopartikler blev derved trukket gennem cellelaget mod magneten. To vitalitets test udført på BCECs, der havde været udsat for SPIOs alene eller suppleret med det eksterne magnetiske felt viste ingen signifikant ændring i vitaliteten. Det blev derfor konkluderet at SPIOs er potentielt potente carriere til hjernen.

I anden del af tesen var formålet at undersøge og sammenligne to forskellige former for *in vitro* BBB, en statisk og en dynamisk. Den statiske model inducerer BCECs til at danne de fleste BBB karakteristika, men mangler evnen til at forme "shear stress", der er en vigtig tætheds promoverende faktor. BCECs dannelse af barriere blev monitoreret ved hjælp af transendothel elektrisk resistans (TEER) måling og celler blev efterfølgende farvet positive for tight junction proteinet zonula occludens 1. Tætheden af barrieren viste sig at øges ved kontakt co-kultur med astrocytter og en yderligere øgning af tætheden blev observeret ved tilsættelse af binyrebarkhormon til mediet.

Den dynamiske *in vitro* BBB model kan inducere shear stress og er derfor en mere kompleks model. Desværre var det ikke muligt at indsamle pålidelige resultater fra den dynamiske model og derfor kunne dannelse af en blod hjerne barriere i denne model ikke undersøges nærmere. Desuden var det heller ikke muligt at sammenligne de to modeller, men det kunne konkluderes at den statiske *in vitro* BBB model på nuværende tidspunkt virker mest stabil.

Acknowledgements

First of all I would like to thank my supervisor, Torben Moos for giving me the opportunity to start on this Ph.D. study and for support and guidance throughout the study. Torben has been very kind to give me great opportunities to evolve my skills as a scientist and teacher. I would also like to thank my co-supervisors Thomas Jensen and Anker Jon for their input to the methods and results obtained during my study.

I would like to thank everybody at the Biomedicine building for the nice times spend in the coffee room at lunch, Monday bread and social events. Especially thanks to Jacek Lichota for guidance, sharing of knowledge and for the nice talks we have had as "office mates". A big thank to Merete Fredsgaard for being the perfect "travel mate", for moral support and for all her help trough out the years. Also Linda Pilgaard is thanked for her input to this thesis, scientific co-operation and moral support.

"Kompetence Fonden" at the University of Aalborg is thanked for financial support.

Last but not least I would like to thank my family and friends for supporting me, not only in good times but also on a cloudy day. Furthermore I would like to thank Palle for always supporting me and believing in me.

List of papers

This Ph.D. thesis is based on the following papers:

Thomsen, L.B., Lichota, J., Kim, K.S., Moos, T. Gene delivery by pullulan derivatives in brain capillary endothelialcells for protein secretion, *J. Controlled Release*, 2011, 115 (1) 45-50.

Thomsen, L.B., Lineman, T. Lichota, J., Kim, K.S., Visser, G., Moos, T. In vitro delivery of superparamagnetic iron oxide nanoparticles through brain endothelial cells (submitted to *Journal of Controlled Release*) .

Thomsen, L.B. and Moos, T. Modeling the Blood-Brain Barrier: Dynamic versus Static in vitro Blood-Brain Barrier Models, (unpublished).

Thomsen, L.B., Larsen, A.B., Lichota, J., Moos, T. Nanoparticle-derived non-viral genetic transfection at the blood-brain barrier to enable neuronal growth factor delivery by secretion from brain endothelium, *Curr Med Chem.*, 2011, 18(2) 3330-3334.

Thomsen, L.B. Lichota, J, Larsen, T. E., Linemann, T., Mortensen, J.H., Nielsen, K.G.D.J., Moos, T. Brain delivery systems via mechanisms independent of receptor mediated endocytosis and adsorptive-mediated endocytosis, *Curr Pharm Biotechnol.*, 2012 (E-pub ahead of print).

1 Introduction

Drug delivery to the brain is restricted by the vascular barriers of the brain. These barriers constitute the blood-brain barrier (BBB) and the blood-cerebrospinal fluid (CSF) barriers (Fig.1). The BBB covers a 1000 times larger area of the capillary surface than that of the blood-CSF barriers, which makes the BBB the main route of interest for systemic drug delivery to the brain [1, 2]. In this thesis, drug delivery to the brain capillary endothelial cells (BCECs) forming the BBB and across the BCECs into the brain interior is the main target. Therefore, the thesis emphasizes the transfections of BCECs and transport into and trough these cells.

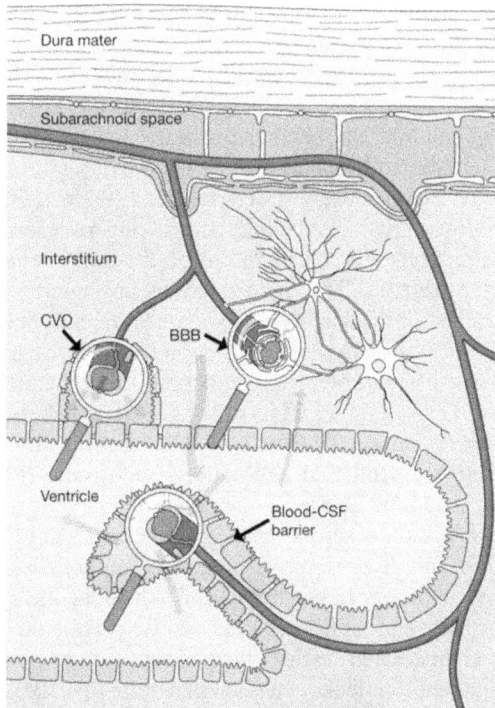

Figure 1 Capillaries in the brain. Cerebral capillaries of blood-brain barrier (BBB), capillaries of the circumventricular organs (CVOs) and capillaries of the blood-cerebrospinal fluid barrier (BCB) in the choroid plexus are highlighted under magnifying glasses. The endothelial cells of the capillaries of the CVO and choroid plexus are fenestrated and leaky and the barrier function is found in the surrounding cells. The capillaries of the BBB are without fenestration and the endothelial cells are interconnected by tight-junctions and they are therefore non-leaky.

1.1   THE BLOOD-BRAIN BARRIER

BCECs denote the BBB and form a major physical restraint on the transport of several molecules present in the blood plasma. Astrocytes and pericytes make intimate contacts with the BCECs and participate in the maintenance of the integrity of the BBB [3]. Astrocytes are especially important for induction and regulation of the BBB properties of the BCECs and their end-feet ensheat almost completely the abluminal surface of the BCECs [4]. Together, the BCECs and astrocytes form a basal lamina present between the BCECs and end-feet of the astrocytes. The basal lamina consists of laminin, type-IV collagen, integrins and fibronectin [5, 6]. The basal lamina is believed to act as a barrier to the passage of macromolecules [2]. The pericytes are embedded in the basal lamina. Studies indicate that pericytes have a role in regulating the paracellular permeability of the BBB by regulating the tight junctions between the endothelial cells [7, 8, 9, 10]. Pericytes have also been shown to be necessary for BBB formation, regulate BBB gene expression, and induce polarization of astrocytic end-feet [9, 10]. The BCECs are also believed to be in direct contact with neurons, suggesting that neurons also could take part in the regulation of permeability of the BBB [11].

The BCECs are thin and non-fenestrated cells [12, 2]. BCECs are rich in mitochondria, hence high metabolic activity, but low in vesicles involved in endocytotic and transcytotic activity [2, 4]. Furthermore BCEC have a higher concentration of drug and nutrient metabolizing enzymes, such as gamma-glutamyl transpeptidase and alkaline phosphatase compared with non-neural endothelial cells [4]. The BCECs are closely interconnected with tight and adherence junctions, which highly impair paracellular trafficking of even small molecules [13, 14]. The tight junctions are considered to be the main structures responsible for the strict barrier properties. The tight junctions are composed of the integral transmembrane proteins occludins, claudins (predominantly claudin 3 and 5), junctional adhesion molecules (JAMs) and endothelial selective adhesion molecule (ESAM). The transmembrane proteins are anchored to the cytoskeleton by zonula occludens 1, 2 and 3 (ZO-1, ZO-2, ZO-3) [3, 4, 6]. The adherence junctions are formed by vascular endothelial cadherins and these are linked to the cytoskeleton by catenins. The platelet endothelial cell adhesion molecule (PECAM) is also a part of the adherence junctions [3, 6].

Transcellular transport across the BCECs takes place by mechanisms like passive diffusion of small lipid soluble, nonpolar compounds; carrier mediated transport of essential nutrients like glucose and amino acids; receptor mediated transport of e.g. insulin and transferrin; adsorptive mediated transport of e.g. albumin; and carrier mediated efflux transporters of amphilic lipid soluble substrates [15, 6]. Furthermore leukocytes can penetrate the BBB transcellularly by diapedesis, giving raise to transvascular transport in the brain [6].

Several lipophilic and cationic drugs which enter BCECs are returned to the plasma by efflux transporters expressed by the BCECs [2]. The entry of large molecules like most drugs into the brain is therefore limited which is additionally supported by the fact that the number of endocytotic and transcytotic vesicles in BCECs are significantly smaller compared with those of capillaries of many other organs [15, 16, 17, 18]. About 98% of all small lipophilic drugs and all large drug molecules with a molecular weight above ~400 Dalton are unable to penetrate the BBB without an enhancing transport strategy [12].

1.2 DRUG DELIVERY TO THE BRAIN

Several strategies have been taken to enable drug transport into the brain via bypass of the BBB impermeability.

Transiently disruption of the BBB integrity can be employed to facilitate entry of drugs to the brain. Disruption is achieved by either opening of tight junctions, by enhancing pinocytotis or by creating lesions in the cell membrane [19, 20]. Disruption can e.g. be mediated by osmotic substances, vasoactive agents, chemicals, and ultrasonic waves. Hyperosmostic substances, such as mannitol cause shrinkage of BCECs and opening of tight-junctions due to an elevation of osmotic pressure [21]. Vasoactive molecules such as bradykinin and histamine are also known to disrupt the BBB [22, 23]. Chemicals like dimethylsulfoxide (DMSO) and ethanol enhance permeation of the BBB by solubilizing the BCECs membrane [19]. Furthermore ultrasonic waves can be employed to create micro-bubbles bursting in the BCECs membrane leading to a higher permeability of the BBB [20]. Administering a drug together with one of these approaches will lead to entry of the drug into the brain through the disrupted areas of the BBB. Unfortunately, not only the drug has access to the brain. The brain is also exposed to e.g. infection, toxins in circulation and plasma proteins. Therefore these procedures can lead to severe damage e.g. serum albumin have damaging effects on astrocytes [19].

Invasive strategies for drug delivery directly to the CNS can also be employed. These delivery methods have the advantage of delivering high drug concentrations directly to the CSF or parenchymal space of the brain and low drug distribution outside CNS [24]. Drugs can be injected by intrathecal catheters in a bolus or continuous infusions [25]. Delivery is high at the site of administration but limited in success due to the poor diffusion of drugs into the brain tissue. Convection-enhanced diffusion (CED) by implanted osmotic pumps increases the distribution of the drug. The diffusion rate is though still not high enough for the drug to reach into the entire brain parenchyma [12]. Intracerebral implants have also shown to lead to controlled release of drugs in the brain. Implants are made of polymeric materials which encapsulate the drug [26]. This strategy is also based on diffusion of the drug from the implant into the brain parenchyma and has the same diffusion limitations as CED [19]. With the invasive delivery strategies follows a risk of increased intracranial pressure due to the increased fluid volume. There is furthermore a higher risk of infection in the brain, because of the need of repeated craniotomy to allow continuous drug infusion [19].

1.2.1 Non-CNS-invasive approaches to enable drug delivery to the brain

Systemic delivery of drugs into the blood-stream for transvascular delivery to the CNS is non-invasive strategies for drug delivery. Drugs are administered through intravenous, intra-arterial or intra-nasal delivery [26]. The delivery bypasses the first-pass metabolism allowing fast access to the brain vasculature [24].

Intra-nasal delivery bypasses the BBB. Due to the highly permeable nasal epithelium, drugs can diffuse across the nasal mucosa, though the arachnoid membrane and into the olfactory CSF compartments [12, 27]. Frequent intra-nasal administration of drugs damages the nasal mucosa and only some drugs, mostly lipid soluble reached into the CSF by this strategy [19].

Intravenous delivery is limited by the non-brain-specific delivery as the drug is circulated throughout the entire vascular system of the body [26].

Intra-arterial delivery is local delivery to the brain as the blood is supplied directly to the brain before entering peripheral tissue. The intra-arterial delivery ensures a higher concentration of drugs delivered to the brain compared with intravenous delivery [28]. Drugs delivered intravenously or intra-arterially for the purpose of entering into the brain are limited by the BBB. If a drug in circulation is to cross the BBB there are great restrictions. To overcome the blood-brain barrier a drug should meet one of the following criteria:

1) Affinity for nutrient transporters or membrane receptors. An example of a substrate that is able to penetrate the BBB by this criterion is a precursor of dopamine, L-3,4 dihydroxyphenylalanine (L-DOPA). L-DOPA is a substrate for the BCEC receptor, large amino acid transporter 1 (LAT1) and is therefore able to cross the BBB without modification. L-DOPA is considered a pro-drug administered to patients with Parkinson's disease to increase dopamine concentration [12].

2) Capability to undergo adsorptive transport e.g. by means of positive charge. Cationic albumin is able to be internalized by BCECs by electrostatic interaction with BCEC membrane proteins [19].

3) Small in size and high lipophilicity. Diazepam, a benzodiazepine, is small (284.7 Da) and highly lipophilic. Diazepam is able to diffuse passively through the BBB [29]. Diazepam is administered e.g. to patients with epileptic seizures or anxiety disorders.

If a drug does not have affinity for BCEC membrane transporters, receptors, or are small and lipophilic, it can be transported by a substance that fulfills these criteria. Drug and gene carriers are such transport vectors that enable or improve delivery of large molecules such as drugs and genetic material to a target organ.

1.3   GENE THERAPY AND DELIVERY TO THE BBB

Gene therapy was first proposed as a treatment of human diseases in 1972 by Fiedmann and Roblin [30]. Expression of disease causing genes can be corrected by gene therapy by the transfer of genetic material into target cells in order to enhance or inhibit production of a protein [31]. Gene inhibitors, such as oligonucleotides and short interfering RNA (siRNA), silence defective genes on the mRNA level in the cell cytosol. Gene enhancers such as complementary DNA (cDNA) compensates for a deficiency in the production of a specific protein [32]. Ideally cDNA is transported into the target cells by a carrier and further into the nucleus where it is integrated into the host cell genome (Fig. 2). If the integration of the cDNA is successful, it will be transcribed and the encoded protein synthesized by the transfected cell [32, 33].

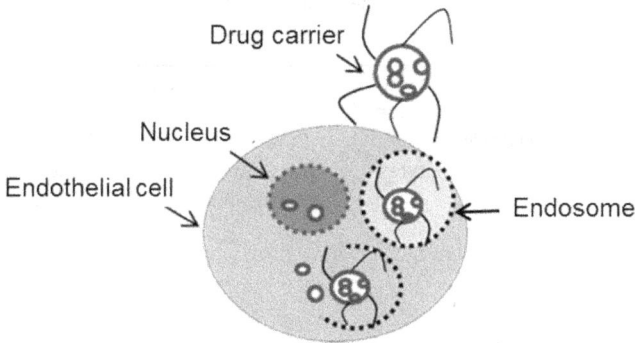

Figure 2 Schematic drawing showing delivery of genetic material to an endothelial cell. A drug-carrier loaded with plasmid cDNA binds to the cell surface, is transported into an endosome from where it escapes and releases its plasmid cDNA into the cytosol. The plasmid DNA may enter the nucleus and incoorperate into the host genome.

Table 1 displays a list of proteins which have been shown to have therapeutic effects on CNS disorders. The proteins in Table 1 are not able to cross the BBB but gene delivery enables delivery of cDNA coding for proteins across the BBB. Introducing cDNA coding for one of the proteins listed in Table 1 to cells of a patient with a CNS disease would lead to production of the protein in the target cells. Genetic material coding for glial cell line-derived neurotropic factor (GDNF) was introduced into mice before dopaminergic nerve damages were induced. GDNF was shown to act as a neuroprotective agent on the dopaminergic neurons [34].

For the introduction of cDNA to a cell it is important to develop an efficient delivery agent, a gene carrier. Naked plasmid cDNA can be delivered systemically without a gene carrier but is rapidly broken down by nucleases and cleared by the mononuclear phagocytic system [35]. Naked genetic material are therefore typically conjugated to the surface or encapsulated inside the core of a gene carrier [33]. Gene carriers can be of viral or non-viral origin. Viral delivery is administered by viral vectors which are based on a natural virus [36]. Non-viral delivery can grossly be divided into physical and chemical approaches [32]. Table 2 displays common viral and non-viral drug-carriers used for delivery of genetic material to the CNS. Physical non-viral gene delivery methods (see table 2) are directly delivered to the cell cytosol whereas the chemical non-viral vectors needs to escape the endosome/lysosome after cellular uptake [32]. Physical non-viral delivery is invasive, difficult to apply and inappropriate for large scale transfection [32]. Gene delivery by viral vectors exploits the natural abilities of viral gene transfer to host cells. Delivery can be successful in both dividing and non-dividing cells as they are able to escape the endosomes and deliver DNA into the cell nucleus [37].

Table 1: Proteins with therapeutic effect in CNS disorders

Protein	Therapeutic effect	CNS disorder	Reference
Brain-derived neurotropic factor (BDNF)	Neuroprotection and neuroregeneration Differention of oligodendroctytes	Multiple Sclerosis Cerebral Ischemia Depression Parkinson's disease	Makar et al 2009 [112] Yong 2009 [124] Zhu et al 2011 [125] Yu and Chen 2011[126] Sun et al 2005 [127]
Glial cell line-derived neurotropic factor (GDNF)	Neuroprotection and neuroregeneration	Parkinson's disease	Biju et al 2010 [34]
Neural growth factor (NGF)	Promote neuronal growth	Alzheimer's disease	Li et al 2008 [128]
Growth hormone (GH)	Neuroregeneration and neuroprotection Proliferation of astrocytes, neurons and oligodendrocytes	Mild cognitive impairment and Alzheimer's disease	Zhang et al 2010 [129] Isgaard et al 2007 [114]
Basic fibroblast growth factor (bFGF)	Neuroprotective	Brain Ischemia	Song et al 2002 [130] Ma et al 2008 [121]
Erytropoetin (EPO)	Neuroprotective and neuroregenerative Oligodendrogenesis	Parkinson's Disease Brain Ischemia/hypoxia	Boado et al 2010 [132] Xue et al 2007 [133] Iwai et al 2010 [134]

Although most viral genetic material is removed from the viral vectors, leaving only sequences for delivery of the exogenous genetic material, there are still concerns about the use of viral vectors [36]. Integration into the host genome has been shown to come with a high risk of insertional mutagenesis. Furthermore the innate immune system are of risk of recognizing surface antigens on the viral vectors which can lead to destruction of all the virally transfected cells [36, 37].

Table 2: Common viral and non-viral gene delivery vectors

Delivery vector	Reference
Viral vectors:	
Retrovirus	Lundberg et al 2008 [135]
Adenovirus	Thaci et al 2011 [37]
Adeno-associated virus	Gray et al 2010 [136]
Herpes virus	Sun et al 2005 [127]
Non-viral vectors:	
Physical techniques:	
Microinjection	Zang and Yu 2008 [137]
Gene gun	Benedicksson et al 2005 [138]
Electroporation	De Vry et al 2010 [139]
Magnetofection	Scherer et al 2002 [77]
Chemical techniques:	
Lipid based:	
Cationic liposomes (Lipoplexes):	
e.g.	Tros de Ilarduya et al 2010 (Review) [140]
Anionic PEGylated immunoliposomes (PILs)	Caveletti et al 2009 [141]
Lipid coated DNA complexes (LCDC)	Skjørringe et al 2009 [123]
Polymer-based:	Lehthinin et al 2008 [142]
Cationic polymers (Polyplexes)	
e.g.: Polyethylenimine	Tros de Ilarduya et al 2010 (Review) [140]
	Son et al 2011 [143]
Dendrimers	
Polymeric micelles	Svenson 2009 [144]
	Shao et al 2010 [145]

Non-viral chemical gene vectors are less efficient than the viral vectors as they often lack natural strategies for endosomal/lysosomal escape and nuclear delivery, but the non-viral chemical gene vectors are less immunogenic and easy to prepare in large scale [33, 38].

1.3.1   Delivery by non-viral chemical gene vectors

Non-viral chemical gene vectors can be conjugated with various targeting molecules to increase BCEC internalization [2, 38]. Targeting the BBB can be achieved by conjugating the non-viral gene vector with a ligand that has affinity for a membrane receptor on the luminal side of BCECs [2]. Transferrin is such a targeting molecule which has affinity for the BCEC transferrin receptor. Unfortunately exogenous transferrin is in direct competition with endogenous transferrin and this limits the possibility of its use for delivery. Anti-receptor antibodies are also widely used to target BCEC receptors and are not in competition with the endogenous proteins. OX26 is a monoclonal anti-rat-transferrin receptor antibody that has been shown to be taken up by BCECs [39, 40]. Unlike transferrin which undergoes transcytosis after receptor binding the monoclonal antibody OX26 has been shown to mainly stay within the BCECs after receptor binding and internalization [41, 42]. OX26 is therefore appropriate for targeting BCECs. The transferrin receptor is not only expressed by BCECs but also by other cell types such as epithelial cells of the intestinal crypts, orthochromatic normoblasts, reticulocytes, trophoblasts cells of the hemochorial type of placenta, Sertoli cells of blood-testis barrier, immature erythroid cells, and hepatocytes [43, 44]. Therefore, targeting the transferrin receptor will possibly not lead to exclusive uptake by BCECs. The uptake by BCECs could though be heightened if the conjugates were administered into the carotid artery and thereby pass the capillaries of the brain early in circulation.

When internalized the non-viral gene vectors are enclosed inside an early endosome which matures to a late endosome and fuses with a lysosome. To avoid degradation in the lysosome the carrier has to escape into the cytosol. Some non-viral carriers for example polyethylenimine (PEI) are capable of escaping the lysosomes by a process called the proton sponge effect: In the acidic lysosome, PEI will bind protons which are pumped in and these are followed by chloride ions and water. Eventually this makes the lysosome swell and burst [45]. After escape from the endosome/lysosome the gene vector needs a rapid trafficking to the nucleus, because DNA is degraded as quickly as within 50-90 min in the cytosol due to nucleases [46]. In dividing cells the nuclear envelope is momentarily open during mitosis, hence allowing transport of DNA into the nucleus. In non-dividing cells DNA reside in the cytosol in between cell divisions and is therefore prone to degradation.

Transport through the nuclear membrane pores is restricted because of a pore diameter of only approximately ~25nm [47]. Most DNA fragments are therefore not able to cross the pores without nuclear trafficking. For gene therapy plasmid DNA can be coupled to a nuclear localization signal (NLS) that enables docking to the nuclear membrane pores and subsequent transport to the nucleus [48, 49]. Coupling NLS to plasmid DNA has been shown to enhance nuclear uptake by 10 to 1000 fold [50, 51]. Overall the optimal characteristics of a non-viral carrier would be that it is biodegradable, non-toxic and have a high delivery rate. Furthermore it should protect its cargo from degradation and be able to deliver its cargo to the cell nucleus.

In this thesis the focus is on two different non-viral carriers, the polyplex: Pullulan-Spermine and superparamagnetic iron oxide nanoparticles (SPIOs) and will be described in further detail in the next sections.

1.3.2   Polyplexes

Polyplexes are complexes consisting of cationic polymers and DNA. Cationic polymers consist of large organic molecules; they include polypeptides, polysaccharides, polyamines and dendrimers.

Many different cationic polymers have been developed for the purpose of polyplex formation e.g. PEI [52], poly-L-lysine (PLL) [53], polysaccharides like chitosan [54], and polyamidoamine dendrimers (PAMAM) [55].

Due to the electrostatic bindings between cationic polymers and anionic DNA, the polymers are excellent carriers of DNA and able to condense DNA to a small size of importance for BBB penetration. An important criterion for the strength of the polymer binding to the DNA is that it has to be sufficiently strong to carry the DNA into the target cell, but at the same time weak enough to allow the separation from the DNA in the cytosol.

The ratio of cationic polymer and DNA in a polyplex is determined by its N/P ratio in where the N refers to the number of nitrogen atoms in the amine groups of the polymer and P to the phosphor content in the DNA. If the polymer contains many branches of amine groups the transfection rate is increased and the toxicity is lowered, e.g. as seen in branched PEI [56].

In circulation, cationic complexes are in risk of being bound to negatively charged albumin, which hinders them from entering the cells. This phenomenon may occur both *in vitro* and *in vivo* [33, 57]. Moreover, when polyplexes are administered intravenously they are often recognized by the immune system as exogenous material and scavenged [33]. Coating the polyplexes with polyethylene glycol (PEG) known as PEGylation shields the polyplexes from this clearance [58, 59].

The positive charge of the polymers enables interaction with anionic glycoproteins and proteoglycans residing on the surface of the cells [60]. Concerning their cellular entry, polyplexes are believed to undergo unspecific cellular uptake by endocytosis [61, 62]. Thereafter, they need to escape the endosomeal/lysosomal system to avoid degradation, which can occur by the so-called proton sponge effect (see above). The proton sponge effect can be created by introducing histidine residues to the polymers [63]. Surface modifications of the polyplexes may also facilitate their escape into the cytosol from the endosomeal/lysosomal system. Hence, PEGylation of the cationic polymers is known to enhance this escape [53].

1.3.3   The novel drug carrier pullan-spermine and gene delivery

Pullulan-spermine is a novel natural cationic complex suitable for forming polyplexes (Fig. 3). Pullulan is a water-soluble extracellular polysaccharide with repeated units of maltotriose condensed trough α-1,6 linkage [64] (Fig. 3). Pullulan is produced by the polymorphic fungus *Aureobasidum pullulans* [65]. Spermine is a polyamine present in all eukaryotic cells and is involved in basic cellular metabolism (Fig. 3). Coupling spermine to a non-viral carrier increases the transfection efficiency [66]. Pullulan is not a cationic molecule but can be cationized by introducing spermine into its hydroxyl groups [64]. Negatively charged plasmid DNA interacts with cationized spermine branches and the more spermine the more DNA is complexed with pullulan-spermine [64].

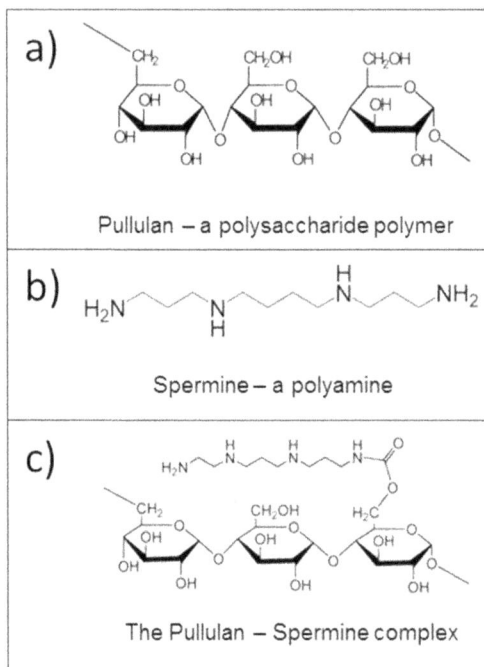

Figure 3 The chemical structures of pullulan, spermine and the pullulan-spermine complex.

Pullulan has affinity for asilaloglycoprotein receptors which is highly expressed by hepatocytes in the liver [65]. However, the pullulan-spermine complex is also internalized in cells that do not express asilaloglycoprotein receptors [65]. Pullulan-spermine is thought to undergo cellular endocytosis both with clathrin or raft/caveolae-dependent endocytosis [65]. It is believed that pullulan-spermine complexes larger than 200nm enters the cells via calveolae-dependent endocytosis and complexes smaller than 200nm are internalized by clathrin-dependent endocytosis [65]. Following internalization of plasmid DNA conjugated with pullulan-spermine, plasmid DNA enters into the nucleus while pullulan-spermine complexes only gets into the cystosol [65]. This polyplex does not have any NLSs suggesting nuclear entrance may occur mainly during mitosis [65].

Pullulan-Spermine has shown good potential as a non-viral carrier of DNA for transfection of various cell types *in vitro* i.e. human bladder cancer cells (T24) [65], human hepatoma cells (HepG2) [64, 67] and mesenchymal stem cells [68, 69].

1.3.4   SPIOs and Blood-Brain Barrier Penetration

A relatively new approach within the field of drug delivery to the brain is the use of magnetic nanoparticles as drug carriers. Magnetic nanoparticles have been applied for diagnostic purposes for about 40 years, but in the last decade their applications have been intensified [70]. They are currently used for many purposes both in basic research and clinical medicine e.g. as a contrast agent for magnetic resonance imaging (MRI) [71], induction of hyperthermia for tumor therapy [72], cell labeling and separation [73, 74], drug delivery [75, 76], and magnetofection [77].

SPIOs are a subtype of SPIOs that is highly magnetizable and have a core of iron-oxide like magnetite (Fe_3O_4) or maghemite (γ-Fe_2O_3) that both are half-metallic. SPIOs have a mean diameter of around 50-100nm [78]. The iron oxide particles show low toxicity and will in time be broken down in the organism to Fe^{2+} and Fe^{3+} that gets incorporated in hemoglobin [78]. SPIOs have been shown to induce oxidative stress in murine macrophage (J774) cells, but only in doses higher than 100μg/ml [79]. For improved visual detection, the magnetic core can be coated with a fluorescent dye. The surface of the SPIOs can furthermore be coated with organic or inorganic substrates e.g. dextran, chitosan, starch, phospholipids or PEG [80, 81, 82, 83]. A coat of PEG can prolong the time in systemic circulation, just as with the polyplexes described above, because they are made less prone to clearance by the mononuclear phagocytic system [76, 83]. Uncoated SPIOs tend to aggregate because of a strong dipole-dipole attraction between the particles. This can be avoided by coating the particles with monomers, inorganic materials or polymers e.g. starch or dextran [80, 84]. A coat of polymeric materials has also been shown to protect the particles from oxidation and thereby making the particles more biocompatible [80, 84]. Furthermore a surface coat of e.g. chitosan or phospholipids enables conjugation of e.g. antibodies, DNA and/or drugs to SPIOs [80, 82, 84, 85].

The SPIOs are also very potent for targeted drug delivery. With the aid of a magnetic force, they are able to very precisely deliver their cargo to a target organ. A magnetic field is supplied by an external magnet or an implanted magnet. When applied, the SPIOs are drawn towards the magnet and concentrated in the area where the magnet is located. The delivery can therefore be very locally and in consequence, fewer particles will be directed towards other non-target areas enabling reduced dosage. The lower dose of nanoparticles will presumably also lead to a reduced risk of unwanted side-effect [79, 80, 85].

1.4   *IN VITRO* BLOOD-BRAIN BARRIER MODELS

Modeling the morphology and permeability of the BBB has been an important issue for decades. The experimental conditions *in vitro* are often more controllable than those *in vivo* and they are overall also more ethically acceptable as the usage of cell lines results in lower use of laboratory animals. Although the BBB formed *in vitro* models lacks the full complexity of the *in vivo* BBB many parameters of the *in vivo* conditions can be assayed *in vitro* e.g. tight junction expression, luminal to abluminal transport of large molecules, and gene expression experiments of the BBB.

A valid real-time monitor of the integrity of the BBB *in vitro* is made by measurements of the trans endothelial electrical resistance (TEER). Unfortunately, BBB in *in vitro* models does not express as high TEER values as can be measured on the BBB *in vivo* [86]. *In vivo* BBB TEER values are in the range of 1200-1900 Ω*cm^2 and have even been measured as high as 8000 Ω*cm^2 [86, 87]. *In vitro* models using cultured endothelial cells generally have a TEER value around 6-10 times lower as those recorded *in vivo* [86].

For *in vitro* studies of the BBB both primary and immortalized cells are being used. BCECs of an *in vitro* BBB model should express as many endothelial markers e.g. ZO-1 and PECAM-1 as possible. Primary BCECs have been isolated and cultured from most mammals with the foremost coming from rat, human and bovine brains (e.g.[88, 89, 90]). The major advantage of primary cells is that they express most of the *in vivo* BBB properties to a higher extent than those of immortalized cells. Most of the immortalized cell lines have been derived from the same species as those of the primary cells and subsequently immortalized e.g. by introducing simian virus 40 (SV40) T antigen. Examples of immortalized cell lines are rat brain endothelial cells, RBE4 and human brain endothelial cells, hCMEC/D3 [91, 92]. Immortalized endothelial cells form less tight BBB properties, which can be seen as a lower TEER values than in vivo or in primary culture, and they do not consistently express endothelial cell markers [93]. Many immortalized cell lines also tend to lose their BBB properties after having been passaged many times in culture [87].

In a model of a well formed BBB BCECs obtain the same polarized properties as can be found in BCECs *in vivo*. The polarized BCECs will form a barrier with an apical membrane facing the lumen of the vessel, a basal membrane facing the abluminal brain side, and a lateral membrane containing tight junction proteins facing the lateral membrane of adjacent BCECs. The various domains of the BCEC membrane have distinctive characteristics determining their function. The mechanisms that induce polarization are not fully understood, but astrocytes are known to secrete a number of substances that participates in the induction of the BBB e.g. basic fibroblast growth factor (bFGF) and angiopoitin 1 (ANG1) [3].

Astrocyte conditioned medium (ACM) have been shown to increase the barrier properties of the endothelial cells [94]. The ACM is obtained from astrocytes in culture and is believed to contain some of these BBB inducible factors like bFGF and ANG1. The *in vitro* BBB model is improved by addition of ACM to the culture media or even better by co-culturing astrocytes with BCECs.

Pericytes are also known to induce a tighter BBB and therefore they may be included in a triple co-culture model for *in vitro* BBB studies together with BCECs and astrocytes [95]. Furthermore, elevation of cAMP in the growth media by addition of hydrocortisone strengthens the BBB properties of BCECs [88, 94, 96].

Optimal properties of an *in vitro* BBB model are reflected in high expression of tight junction proteins that lead to an accordingly high TEER value, expression of BBB transporters, and in low tracer permeability of e.g. sodium fluorescein or sucrose.

1.4.1   Static *in vitro* Blood-Brain Barrier Model

Static *in vitro* models have been employed for decades (e.g. [88, 95, 97, 98]). They are based on the insertion of a microporous membrane filter into the well of a culture plate (Fig. 4). Brain endothelial cells are cultured on the membrane in the insert, hence forming a monolayer which models an intact BBB.

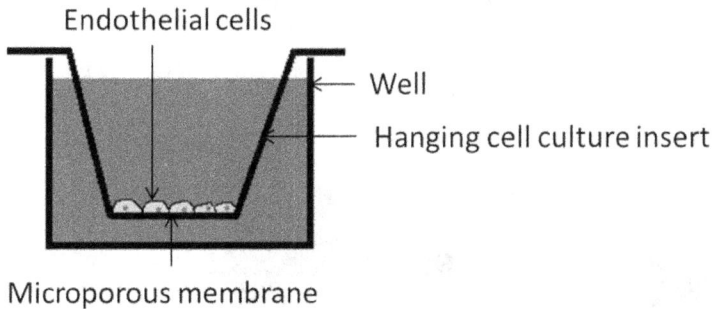

Figure 4 A hanging cell culture insert inserted into a well of a culture plate. A microporous membrane forms the bottom of the insert. In the insert BCECs can be cultured in a monolayer.

Astrocytes can be cultured in the well underneath the inserts, which corresponds to the abluminal side of the BBB, in a non-contact co-culture. The astrocytes can also be seeded on the membrane on the outside of the inserts, which ensures direct contact between the astrocytes and endothelial cells through the microporous membrane and is therefore called a contact co-culture. In both co-culture forms the TEER can be measured with the aid of two electrodes. One electrode is inserted into the well and the other into the insert. The two electrodes are separated by the endothelial cell layer and its electrical resistance is measured. TEER measurements of the static *in vitro* BBB models are generally lower than *in vivo* conditions although a few studies have reported on high TEER values reaching those of TEER values *in vivo* [87].

1.4.2   Dynamic *in vitro* Blood-Brain Barrier Model

Dynamic *in vitro* BBB models are based on the fact that they are able to create shear stress. Shear stress is able to induce most features of the BBB phenotype e.g.

BBB tightness. *In vivo* shear stress is the mechanical pressure generated by the blood flow exerted on the luminal surface of the endothelial cells [99, 100]. Shear stress is not created in the static *in vitro* BBB models.

In this thesis the emphasis will be on the dynamic *in vitro* blood-brain barrier model "Flocel" made by Flocel.Inc, USA [100, 101, 102, 103, 104, 105, 106, 107, 108, 109]. The Flocel model (Fig. 5) consists of a cartridge with two compartments.

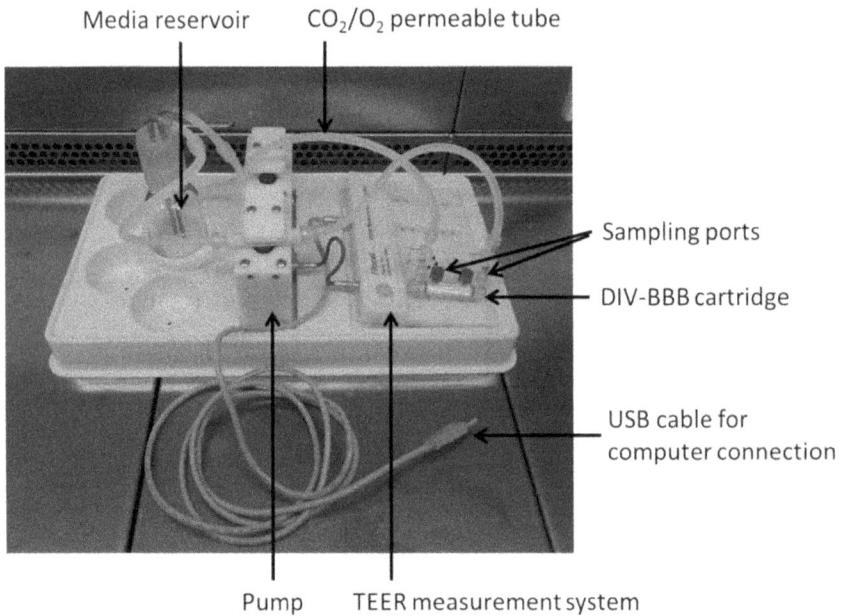

Figure 5 The dynamic in vitro BBB model. The model consists of a cartridge placed in a TEER measurement system. The DIV-BBB cartridge has an inner compartment consisting of 19 hollow fibers made of a microporous membrane. The cartridge has four samplings port and four electrodes for measuring

The inner compartment is made up of hollow fibers which mimic brain capillaries. On the outside of the hollow fibers, the outer compartment constitutes the surrounding space mimicking the brain extracellular space. The cell media is pumped via CO_2/O_2 permeable tubing through the hollow fibers creating shear stress along the inner surface. Sampling ports connect to both the inner and outer chambers from where media can be collected. The bottom side of the cartridge contains four electrodes that allow for measurement of TEER values of the cells placed inside the hollow fibers to form a barrier. BCECs can be seeded in the inner chamber of the hollow fibers. Furthermore astrocytes can be seeded in the outer chamber at where they can grow to cover the entire abluminal side of the hollow fibers and form direct contacts with BCECs through micro pores in the fiber walls. TEER values in the Flocel model have been measured to values around 1200 $\Omega*cm^2$, which is near the TEER of the BBB *in vivo* (e.g. [109]).

2 Objective of the thesis

Over all the objective of this study is to find applicable drug carriers for delivery to BCECs. Furthermore the objective is to establish the best possible *in vitro* BBB model for testing the application of drug carriers. For further description the objective can be divided into three separate aims for further description.

1) The first aim is to investigate a novel non-viral carrier pullulan-spermine for its abilities to function as a transfection agent at the BBB. Pullulan-spermine has been proved to be able to carry cDNA into various cell types and therefore this part of the thesis aims at exploring, if the carrier also has capabilities of gene delivery to BCECs. If pullulan-spermine could successfully deliver DNA to the BCECs then it would be interesting to detect whether the DNA also would be transcribed and expressed by the BCECs. It would also be interesting to investigate whether the BCECs would be able to produce and secrete the DNA encoded protein.

2) SPIOs can potentially be used for targeted delivery and the second aim of the thesis is to investigate if SPIOs would be able to enter into and cross the brain capillary endothelial cells. This would involve the application of an external magnetic force that can pull the SPIOs towards the source of the magnetic field. The particles could therefore potentially be very precisely delivered. Therefore the aim in this part of the thesis is to test the ability of magnetic particles to pass through BCECs cultured in an *in vitro* BBB model with and without an external magnetic source. The impact of SPIOs and the external magnetic source on BBB integrity and BCEC vitality is also investigated.

3) The third aim is to characterize a new dynamic *in vitro* model of the BBB, which can be used for testing the ability of various drug and gene carriers to penetrate the BBB. The model has been claimed to exceed the abilities of other models in the field to replicate the BBB. The dynamic model will therefore be compared with a well-established static model. A good model should be able to express as many BBB characteristics as possible and therefore give reasonable indications of the abilities of the carriers to penetrate the *in vivo* BBB.

3 Results

3.1   STUDY I

GENE DELIVERY BY PULLULAN DERIVATIVES IN BRAIN CAPILLARY ENDOTHELIAL CELLS FOR PROTEIN SECRETION

Louiza Bohn Thomsen, Jacek Lichota, Kwang Sik Kim and Torben Moos

The manuscript was published in Journal of Controlled Release, Vol. 515, Issue 1, 45-50, 2011.

Journal of Controlled Release 151 (2011) 45–50

Contents lists available at ScienceDirect

Journal of Controlled Release

journal homepage: www.elsevier.com/locate/jconrel

GENE DELIVERY

Gene delivery by pullulan derivatives in brain capillary endothelial cells for protein secretion

Louiza Bohn Thomsen [a],[*],[1], Jacek Lichota [a],[1], Kwang Sik Kim [b], Torben Moos [a]

[a] Department of Health Science and Technology, Biomedicine, Aalborg University, Fredrik Bajers Vej 3B, 9000 Aalborg, Denmark
[b] Division of Pediatric Infectious Diseases, Johns Hopkins University School of Medicine, 160 North Wolfe Street, Park 256, Baltimore, MD 21287, USA

ARTICLE INFO

Article history:
Received 2 July 2010
Accepted 4 January 2011
Available online 18 January 2011

Keywords:
Blood-brain barrier
Pullulan
Gene therapy
Gene expression
Spermine

ABSTRACT

The blood-brain barrier (BBB) formed by brain capillary endothelial cells protects the brain against potentially harmful substances present in the circulation, but also restricts exogenous substances such as pharmacologically acting drugs or proteins from entering the brain. A novel and rather unchallenged approach to allow proteins to enter the brain is gene therapy based on delivery of genetic material into brain capillary endothelial cells. In theory in vivo transfection will allow protein expression and secretion from brain capillary endothelial cells and further into the brain. This would denote a new paradigm for therapy to transport proteins across the BBB. The aim of this study was to investigate the possibility to use brain capillary endothelial cells as factories for recombinant protein production. Non-viral gene carriers were prepared from pullulan, a polysaccharide, and spermine, a naturally occurring polyamine that were additionally conjugated with plasmid DNA. We were able to transfect rat brain endothelial cells (RBE4) and human brain microvascular endothelial cells (HBMECs). Transfection of HBMECs with pullulan-spermine conjugated with plasmid DNA bearing cDNA encoding human growth hormone 1 (hGH1), led to secretion of hGH1 protein into the growth medium. Hence, the pullulan-spermine delivery system is a very promising method for delivering DNA to brain endothelial cells with potential for using these cells as factories for secretion of proteins.

© 2011 Elsevier B.V. All rights reserved.

1. Introduction

The blood-brain barrier (BBB) is made of non-fenestrated brain capillary endothelial cells connected by tight junctions that restrict paracellular diffusion of solutes or drugs into the brain [1]. Influx and efflux transporter mechanisms exclusively for specific molecules in the brain capillary endothelial cells make up a barrier for transcellular transport of the most exogenous molecules including polypeptides [2,3]. Polypeptides such as brain-derived neurotrophic factor (BDNF), erythropoietin (EPO), growth hormone (GH) and fibroblast growth factor (FGF) have been proved to have neuroprotective and neuroregenerative effects [4–11]. These neurotrophic agents would therefore be useful in the treatment of CNS injuries and disorders. However, these macromolecules are prevented from entering the brain in adequate amounts to exert their therapeutic effect [4–11]. A novel and rather unchallenged

approach to allow polypeptides to enter the brain is gene therapy based on delivery of genetic material into brain capillary endothelial cells (BCECs). The in vivo transfection will allow protein secretion from BCECs and further into the brain and denotes a new paradigm for therapy to transport polypeptides across the BBB [12]. There are two general classes of carriers for gene delivery, i.e. viral and non-viral [13]. Viral carriers are often highly efficient, but they are connected with a risk of not being biologically safe. The non-viral carriers have the advantage of being biologically safe, they exhibit low cytotoxicity, are easy to prepare and can carry large DNA fragments [14].

Pullulan-spermine complexed with plasmid DNA has formerly been shown to be a potent carrier system for non-viral gene therapy [14–21]. Pullulan is a water soluble polysaccharide [14,16,17,22]. Spermine is a naturally occurring polyamine present in all eukaryotic cells and is involved in basic cellular metabolism [14,17,20]. Pullulan-spermine is known to undergo cellular endocytosis via clathrin or raft/caveolae dependent endocytosis, but the mechanisms leading to its cellular uptake and further internalization are not known. Pullulan might be recognized by the asialoglycoprotein receptor (ASGPR) found primarily in the liver, but cells which do not express ASGPR are also able to internalize pullulan-spermine [16].

In this study, we were able to construct both cationic and anionic pullulan-spermine complexes and demonstrate the capability of the cationic complexes to transfect immortalized human brain microvascular endothelial cells (HBMECs) and rat brain endothelial cells (RBE4s)

Abbreviations: BBB, blood-brain barrier; RBE4, rat brain endothelial cells 4; HBMEC, human brain microvascular endothelial cells; hGH1, human growth hormone 1; ASGPR, asialoglycoprotein receptor; PICs, polyion complexes; BDNF, brain-derived neurotrophic factor; FGF, fibroblast growth factor; EPO, erythropoietin.
* Corresponding author. Section of Neurobiology, Biomedicine, Department of Health Science and Technology, Frederik Bajers Vej 3B, Aalborg University, DK-9220 Aalborg East, Denmark. Tel.: +45 99407461; fax: +45 96357816.
E-mail address: lbt@hst.aau.dk (L.B. Thomsen).
[1] These authors contributed equally to this work.

GENE DELIVERY

using plasmid DNA containing cDNA for HcRed fluorescent protein as a reporter gene. We prove the secretion of human growth hormone 1 (hGH1) after transfection of HBMECs, hence providing evidence for a novel carrier for non-viral gene therapy to BCECs followed by protein secretion.

2. Materials and methods

2.1. Materials

Pullulan with an average molecular weight of 68.9 kDa was purchased from Hayashibara Biochemical Laboratories, Inc. Okayama, Japan. Spermine was purchased from Applichem, Darmstadt, Germany. Other chemicals were obtained from Sigma-Aldrich (Germany) and Applichem and used without further purification.

2.2. Preparation of pullulan derivatives

25 mg of pullulan was dissolved in 2.5 mL dimethyl-sulfoxide (DMSO). 435 mg of carbonyldiimidazole (CDI) was added to pullulan in DMSO to reach a molar ratio of 3:1 for CDI to the hydroxyl groups of pullulan. The CDI-pullulan mixture was incubated for 5 min at room temperature. 2.5 g spermine dissolved in 22.5 mL DMSO was added drop-wise to reach a large molar excess of spermine in order to prevent cross linking of pullulan. The mixture was then incubated overnight at 40 °C. The conjugates were purified by dialysis into MilliQ water in dialysis sacs with a molecular cut off weight of 10 kDa. The pullulan-spermine conjugates were used immediately after dialysis or stored at 4 °C.

2.3. Preparation of plasmid DNA

The following plasmids were used: pHcRed1-C1 (Clontech, USA), pCMV6Entry-GH1 (Origene, USA). Both plasmids were transformed by heat-shock into chemically competent E. coli strain DH5α (Invitrogen) and purified by ion exchange chromatography with Nucleobond Xtra Midi (Macherey-Nagel, Germany) according to manufacturer's protocol.

2.4. Electrophoresis of spermine-pullulan-plasmid DNA complexes

Polyion complexes (PICs) of DNA and spermine-pullulan were prepared in 10 mM PBS solution at different DNA/spermine-pullulan w/w ratios. After 15 min of incubation, the complexes were mixed with loading buffer (0.1% sodium dodecyl sulfate, 5% glycerol, and 0.005% bromophenol blue) and run on 0.8% agarose gel in Tris-acetate-ethylenediaminetetraacetic acid buffer solution (TAE) containing 0.1 mg/mL ethidium bromide (EtBr) at 100 V for 30 min. The gel was imaged with Kodak Image Station 4000MM Pro (Carestreamhealth, USA).

2.5. Characterisation of PICs

Size (DLS/Non-Invasive Back-scatter (NIBS)) equivalent to particle diameter and charge (ζ-potential) were measured on Zetasizer Nano (Malvern, UK). The PICs were prepared in 10 mM PBS solution at different DNA/spermine-pullulan ratios. The size of PICs was analyzed based on the Cumulants method by computer software. The Rs value was calculated automatically by the equipped computer software and expressed as the apparent molecular size of samples. The ζ-potentials were automatically calculated by the software. The measurements were done at least three times for every sample.

The total nitrogen content of the pullulan–spermine complexes was measured in agreement with the Danish Standard (DS 221). In short the pullulan–spermine complex was oxidized in a solution of $K_2S_2O_8$, NaOH and double-distilled water and boiled for 30 min at

200 kPa and 120 °C. The pH was adjusted with H_2SO_4 to a pH value of 7 and the nitrogen concentration was measured on an autoanalyser (Technicon Traacs 800, Technicon Instruments, Pakistan) and afterwards the nitrogen content was calculated. Phosphorous content in DNA was set to 9.4%. From these values the N/P ratio was calculated.

2.6. Cell cultures

HBMECs were isolated from a brain biopsy of an adult female with epilepsy and immortalized as described in Greiffenberg et al. [23] and cultured in Medium 199 with L-glutamine and HEPES (Invitrogen) with 10% Fetal Calf Serum (FCS) (Gibco, Invitrogen, UK), 10% Nu Serum IV (BD Biosciences, San Jose, California, USA), 100 U/mL Penicillin G Sodium and 100 μg/mL Streptomycin Sulfate (Gibco). Immortalized RBE4s were cultured in 50% Alpha-MEM with Glutamax-1 (Gibco) and 50% HAM's F-10 (Gibco) with 10% Fetal Calf Serum, 100 U/mL Penicillin G Sodium (Gibco), 100 μg/mL Streptomycin Sulfate (Gibco), 300 μg/mL Geneticin Sulfate (Invitrogen) and 1 ng/mL basic Fibroblast Growth Factor (Invitrogen). All surfaces for culturing RBE4 cells were coated with 3.0 mg/mL bovine collagen type 1 in 0.012 M HCl (BD Biosciences, USA). The cells were seeded in 6 well cluster plate wells (Sarstedt, Nümbrecht, Germany) at a density of 3×10^5 HBMEC cells/well and 4×10^5 RBE4 cells/well in 2 mL culture media. The cells were then cultured in an incubator with 5% CO_2 at 37 °C for 24 h to reach confluency of approximately 60–80% or for approximately 48 h to reach a confluency of 100%.

2.7. In vitro transfection of brain endothelial cells

The PICs and plasmid cDNA were formed according to the protocol of Jo et al. [14]. 10 μl. of pullulan-spermine was added to 40 μL of MilliQ water in one micro tube and 5 μg pCMV6Entry-GH1 or pHcRed1-C1 was added to phosphate buffered saline (PBS) (8.9 g/L NaCl, 0.2 g/L KCl, 5.24 g/L NaH₂PO₄, and 24.024 g/L Na₂HPO₄ dissolved in double distilled water to a final volume of 1 L) with a total volume of 50 μl. in another micro tube. The two solutions were then mixed and left at room temperature for 15–20 min to form cationic complexes. A commercially available transfection reagent Turbofect™ (Fermentas, Lithuania) was used as a control method. First 200 μl. of culture media without FCS was mixed with 2 μg pCMV6Entry-GH1 or pHcRed1-C1 and then 4 μl of Turbofect™ was added. The solution was left at room temperature for 15–20 min to form complexes. The amounts for both the PICs and the Turbofect™ solution are given per well in a 6 well cluster plate. The PICs containing pHcRed1-C1 were tested on both HBMECs and RBE4s whereas the PICs containing pCMV6Entry-GH1 were only tested on HBMECs.

Prior to the addition of the PICs, the media in the wells were exchanged to the media without FCS and penicillin. In some cases the media were exchanged with media containing FCS but without penicillin. In the wells where Turbofect™ was to be added, the media were exchanged to the media containing FCS but without penicillin. In some of the wells no transfection reagents were added and those wells were treated as a negative control whereas in two wells 2.5 ng/well FLAG–BAP fusion protein was added as a positive control for immunoprecipitation assay and as a control for a possible proteolysis in the culture media. 100 μl. of the PICs or the media containing Turbofect and plasmid DNA were added to each well in droplets that were dispersed throughout the wells. The cells were kept in room temperature for approximately 15 min and further cultured for 6 h in an incubator at 37 °C with 5% CO_2. Then, the media in all the wells containing PICs were exchanged with media containing FCS but without penicillin. The cells were put back into the incubator and cultured for 36–48 h. The cells transfected with pHcRed were examined in a fluorescence microscope (Axiovert 200, Carl-Zeiss, Germany). Afterwards the media was aspirated and transferred to centrifuge tubes. The cells were rinsed three times with PBS and both tubes and well plates were subsequently

L.B. Thomsen et al. / Journal of Controlled Release 151 (2011) 45–50 47

stored for short term at −20 °C for further use in an immunoprecipitation assay. Some of the wells with transfected cells, either transfected with pullulan–spermine or Turbofect™ conjugated with pHc-Red1, were trypsinised, stained with Trypan Blue (Sigma-Aldrich) and viable cells were counted.

2.8. Immunoprecipitation, gel-electrophoresis and Western blotting isolation and detection of FLAG-tagged cDNA

To isolate the flag tagged human growth hormone that was secreted from the transfected cells and into the media, a FLAG® Tagged Protein Immunoprecipitation Kit (Sigma-Aldrich) was used according to the manufacturer's protocol. In short, anti-FLAG Affinity resin was washed; first to remove the storage glycerol and secondly to remove unbound anti-FLAG antibody from the resin suspension. Then the resin was incubated with the media from the transfected cells, negative and positive control cells and controls for the assay. All reactions were left overnight at 4 °C with agitation. The resins were then washed and the FLAG-tagged protein was eluted from the resins with SDS PAGE Sample Buffer and run on a 4–12% NuPAGE®-Bis–Tris-minigel in NuPAGE MOPS SDS Running Buffer (Invitrogen) and transferred to a PVDF-membrane with the iBlot®Dry Blotting System (Invitrogen). ProteoQwest™ FLAG® Colorimetric Western Blotting Kit (Sigma-Aldrich) was used as a detection tool for Western blotting. Briefly, the membrane was blocked in PBS containing 3% non-fat milk for 1 h and then incubated for 1 h with the anti-FLAG M2 monoclonal antibody–peroxidase conjugate in PBS containing 3% non-fat milk. Thereafter the membrane was washed three times and incubated with 3,3′,5,5′-tetramethylbenzidine (TMB) solution for approximately 20–30 min until the protein bands were clearly visible. The membrane was washed with double distilled water, dried on blotting paper, scanned and stored in the dark.

2.9. RT-PCR

RNA was purified using Nucleospin RNA II kit (Macherey-Nagel) followed by cDNA synthesis using iScript kit (BioRad, USA). 1 μg of total RNA was used for every first strand cDNA synthesis with random hexamer primers. PCR was performed using DreamTaq Green master mix (Fermentas), 1 μl of each cDNA sample and 10 pmol of each of the primers (Table 1). PCR was run as follows: 1 cycle 95 °C (2 min), 30 cycles 95 °C (30 s), 55 °C (30 s), and 72 °C (30 s), and 1 cycle 72 °C (5 min). The PCR products were run for 20 min at 120 V on a 2% agarose gel containing 0.5 μg/ml ethidium bromide and visualized on the Kodak Image Station 4000MM Pro (Carestreamhealth).

3. Results and discussion

An efficient non-viral gene delivery depends on carriers that can electrostatically bind to a genetic material. The present study demonstrates in vitro gene expression after transient transfection of plasmid DNA by pullulan–spermine derivative into RBE4 and HBMEC

cells. The resulting transfection with plasmid DNA led to secretion of growth hormone protein into the medium, which is a proof-of-concept study opening possibilities for further investigation on transport of proteins into the brain by surpassing the permeability restraints of the BBB [12].

3.1. Preparation and characterisation of PICs

Complexation of plasmid DNA with spermine–pullulan derivatives was chosen as a method to manufacture a carrier for gene delivery into brain endothelial cells in vitro. The spermine–pullulan derivatives were prepared as described in the literature [14]. The PICs of pullulan- spermine and plasmid DNA were run on an agarose gel in order to prove efficient complexation of DNA with pullulan derivatives at various DNA to pullulan w:w ratios (Fig. 1).

After electrophoresis and visualization it has been concluded that PICs have not migrated into the agarose gel due to their size. Instead, the complexes were collected at the edge of each well. It is interesting to note that PICs migrate towards the cathode or anode depending on the amount of DNA complexed, hence the total charge of PICs (Fig. 1), which demonstrates that both catio- and anioplexes have been formed. In order to confirm this observation the ζ-potential of PICs were measured. The measurements have indicated that catioplexes had a positive charge of approximately +10 mV, whereas the anioplexes were approximately −20 mV (Fig. 2b). These results are in good agreement with previous reports describing both types of PICs where catioplexes were measured to be approximately +10 mV, whereas anioplexes were very negatively charged reaching −42 mV [20]. To further characterize the PICs, DLS measurements (a non-invasive back-scatter analysis) were performed. The size of the pullulan-spermine derivative was very difficult to measure and resulted in several peaks suggesting an irregular shape of the conjugates. Interestingly however, both a regular form and compaction were observed upon addition of DNA (Fig. 2a). The catioplexes were usually small in size (approximately 300 nm) whereas anioplexes were reaching a size of 800–900 nm (Fig. 2a) as they were complexed with approximately 10× more DNA. Similar PIC sizes were reported previously for the catioplexes [14] whereas smaller size (349 nm) was reported for anioplexes [20]. The total N content was measured for the anioplexes and catioplexes and the N:P ratios were calculated to be 4 and 2 respectively. Based on the Zetasizer measurements, a fraction of catioplexes were selected and exclusively used for the remaining experiments. These catioplexes had an N:P ratio of 3.4.

Table 1
Primers used for RT-PCR.

hGH1	Forward primer	5′-GTCTATTCGAGCACCCTCCA-3′
	Reverse primer	5′-GGATGCCTTCCTCTAGGTCC-3′
Human GAPDH	Forward primer	5′-CCTCCAAGGAGTAAGACCCC-3′
	Reverse primer	5′-TGTGAGGAGGGGAGATTCAG-3′
Rat tubulin	Forward primer	5′-TAGAACCTTCCTTCCGGTCGT-3′
	Reverse primer	5′-TTTCTTCTCGGGCTGGTCTC-3′
Hc-red	Forward primer	5′-ATGTACATGGAGGGCACCGTGAA-3′
	Reverse primer	5′-AAGCTCTGCTTGAAGAAGTCGGGGAT-3′

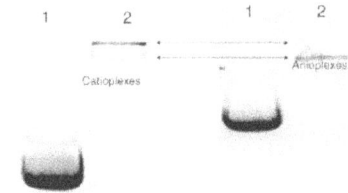

Fig. 1. Agarose gel electrophoresis of pullulan–spermine–plasmid complexes. 1 - free plasmid, 2 - pullulan–spermine–plasmid complexes. The complexes stay in the wells and migrate towards the cathode or anode depending on the charge (anioplexes and catioplexes).

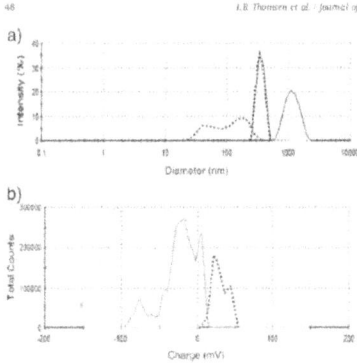

Fig. 2. Measurement of size and charge of PICs. a) Size of plasmid DNA alone (dotted), catioplexes (hatched) and anioplexes (line). b) ζ-potentials of complexes (line) and catioplexes (dotted).

3.2. In vitro transfection of brain capillary endothelial cells

pHcRed1-C1 plasmid containing cDNA for the red fluorescent protein HcRed as a reporter gene was complexed with pullulan-spermine to form cationic complexes. These complexes were made to visualize the transfection ability and distribution of the protein in the endothelial cells. 36–48 h after the cells had been transfected they were observed using fluorescence microscopy (Fig. 3). RBE4s and HBMECs which were transfected at a confluency reaching 60–80% expressed the HcRed protein that distributed to both cytoplasm and nucleus, which is consistent with Kanatani et al. [16]. Despite the fact

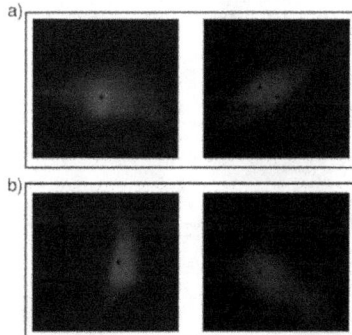

Fig. 3. Pullulan transfected brain capillary endothelial cells expressing Hc-Red. RBE4s (a) and HBMECs (b) were transfected with pullulan–spermine conjugated with pHc-Red1-C1, a red fluorescent reporter gene. Note that the cells express Hc-red in both the cell cytosol and nucleus (*). The cells were observed in a fluorescence microscope with 40× magnification.

that a different assay was used, in which plasmid DNA was labelled with rhodamine, a similar distribution was observed in human bladder cancer cells (T24) [16]. It is believed that after internalization, pullulan-spermine remains in the cytoplasm, whereas plasmid DNA enters the nucleus, most likely under mitosis, when the nuclear membrane transiently disappears [14,16]. The expression of HcRed from the plasmid DNA in RBE4 and HBMEC cells confirms its delivery into cells and supports the hypothesis that plasmid DNA does reach the nucleus. Furthermore we have investigated whether transfection could occur in confluent, non-dividing cells. We have observed that HcRed was also expressed in non-dividing HBMECs (data not shown) thus concluding that plasmid DNA reaches into the nucleus in dividing (60–80% confluency) as well as non-dividing cells (100% confluency) which suggest that internalization of the complexes leads to DNA transfer into the nucleus without necessity of mitosis. This is of great importance in vivo as the brain endothelial cells are non-dividing cells and therefore would not be able to take up plasmid DNA into the nucleus if the transfection depended on a mitotic division.

Consistent with previous findings it was observed that the pullulan-spermine-plasmid DNA complex had high transfection efficiency [16]. Both RBE4s and HBMECs were transfected in parallel with the commercially available transfection reagent Turbofect™ using pHcRed1-C1 as a positive control for pullulan mediated transfection (data not shown). Transfection efficiency similar to the commercial transfection agent Turbofect™ was concluded based on the number of transfected cells present when observed in the fluorescent microscope. The cell viability upon transfection with the PICs was 35% higher than the viability of the cells upon transfection with Turbofect™. The latter confirms the advantage of the PICs in their lack of cellular toxicity, which together with a high transfection rate shows that the PICs are potent transfection agents with low cytotoxicity and high biological safety due to their organic origin. The PICs are therefore suitable gene-carriers for gene delivery to BCECs.

In the above mentioned experiments the PICs were added to the wells containing cells and media without fetal calf serum (FCS) and penicillin, following the protocol from Jo et al. [14]. In in vivo experiments the pullulan-spermine complexes would not be in serum free conditions and therefore the effect of serum on the PICs was investigated. When PICs were added to the cells cultured in media containing serum, considerably lower amount of cells were expressing the red fluorescent protein and the transfection efficiency did not change by changing the amount of pullulan-spermine and pHc-Red1-C1 concentration. This suggests that before the pullulan-spermine complexes can be used as carrier complexes in vivo some alterations must be made to increase the transfection efficiency in the blood environment.

3.3. In vitro transfection of human brain capillary endothelial cells and protein secretion

The culture media was collected 48 h after transfection of the HBMECs and FLAG-immunoprecipitation was carried out in order to purify FLAG-tagged hGH1 if secreted into the medium. The FLAG affinity-bound proteins were then run on an SDS-gel and detected by immunoblotting using an anti-FLAG antibody and visualized using TMB (Fig. 4a). FLAG-tagged-GH1 was detected indicating that the protein was secreted into the culture media after transfection of the HBMECs (Fig. 4a). We conclude that pCMV6Entry-GH1-pullulan complex enters the cell where hGH1 is synthesized and secreted from the HBMECs into the surroundings. The endothelial cells were cultured in a monolayer in a 6-well culture plate and we were therefore not able to define a basolateral and an apical side of the layer. As many substances are transported back out of the endothelial cells by efflux transporters to the apical side we still have to prove that this is not the case for the proteins synthesized by the endothelial cells remaining brain [24]. GH promotes growth, tissue repair and cell regeneration and has been proved to have a neuroprotective effect. Therefore it can be used for treatment of CNS injuries [6,7]. The secretion property of hGH1 and ease of FLAG-tag immunoprecipitation and detection were employed in this study. The pullulan-spermine complex delivered plasmid with cDNA encoding hGH1 into the monoculture of brain endothelial cells thereby making them transient transgenic factories for production and secretion of hGH1. Genes encoding BDNF, FGF and EPO could be assayed for their ability to transfect the brain capillary endothelial cells using the pullulan-spermine conjugate technique and evaluated for their ability to act as a neuroprotectant in disease models.

For in vivo purposes, the delivery of PICs containing genetic material encoding these growth factors, targeted delivery to the brain capillary endothelial cells are likely needed. We are in the process of building up a pullulan-spermine-cDNA complex for in vivo targeting by encapsulating it with a lipid coat that can be attached to an antibody that targets the transferrin receptor. The capillary endothelial cells of the CNS are the only endothelial cells of the body to contain transferrin receptors [25], and this receptor is a preferable target for the delivery of carrier complexes intended for non-viral gene therapy to the brain using the vascular route [12]. The anti-rat transferrin receptor antibody gets internalized by BCECs in vivo and therefore represents a resourceful tool for in vivo targeting of pullulan-spermine complexes [26–29].

4. Conclusions

In this study anionic and cationic pullulan-spermine/DNA complexes were prepared and characterized. Catioplexes had a positive charge of approximately +10 mV and had a size of approximately 300 nm whereas the anioplexes were approximately −20 mV in charge and 800–900 nm in size. HBMECs and RBE4s were successfully transfected with the fluorescent reporter gene pHcRed1-C1 with good transfection efficiency and low cytotoxicity. Secretion of hGH1 protein was detected after in vitro transfection of HBMECs with pullulan-spermine complexed with pCMV6Entry-GH1. We conclude that the pullulan-spermine delivery system is a very promising method for delivering DNA to the brain endothelial cells and for using these cells as factories for protein secretion.

Acknowledgements

We would like to thank Merete Fredsgaard for technical assistance in the laboratory and Professor Thomas Jensen for his input in the project. The work has been kindly supported by the Danish Medical Research Council (Grant no. 271-06-0211), the Spar Nord Fund, and the Obelske Family Fund.

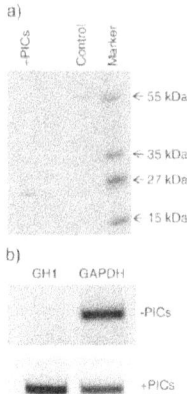

Fig. 4. Expression of human growth hormone in HBMECs. a) Detection of FLAG-tagged hGH1 on a PVDF membrane with TMB. The first lane (+PICs) shows immunoprecipitated FLAG-tagged hGH1 proteins from pullulan transfected HBMECs. A band is seen just below 27 kDa, corresponding to the FLAG-tagged hGH1 which has a size of 21 kDa. The second lane (−PICs) shows immunoprecipitate from non-transfected HBMECs and no band is seen. The third (control) immunoprecipitated FLAG-BAP fusion protein which normally migrates as a 49-55 kDa band. The fourth lane (marker) is a prestained protein ladder. b) RT-PCR analysis of human hGH1 expression after pullulan transfection of HBMECs. hGH1 transcripts were clearly present in transfected HBMECs (+PICs), whereas in non-transfected HBMECs (−PICs), a vague amount of hGH1 transcripts was seen. As a control both non-transfected and transfected HBMECs were shown to express GAPDH.

after transfection with pullulan-spermine, before the method can be used for in vivo studies.

In order to further characterize hGH1 expression, the total RNA was purified from both transfected and non-transfected cells, cDNA was synthesized and RT-PCR performed (Fig. 4b). The PCR reaction confirmed the presence of a high amount of hGH1 transcript in the transfected cells. Another interesting observation has been made concerning expression of hGH1 in non-transfected cells: even though endothelial cells probably are not a main source of hGH1 in vivo, a small amount of hGH1 transcript was detected in the control cells, which is in agreement with previous studies [24]. Expression of a transgene has previously been proved by Thakor et al. [20] after transfections of DRGs, rat dorsal root ganglions, with pullulan-spermine complexed with plasmids expressing human hepatocyte growth factor (HGF). A significant increase in the outgrowth of neurites was seen in dorsal root ganglion cells in the rat after the transfection [20].

In this study we wanted to investigate the possibility to use BCECs as small factories for recombinant protein production. An immunoprecipitation assay was developed for detection of secretion of a given protein directly into culture media after pullulan-spermine mediated transfection. pCMV6Entry-GH1 plasmid was used bearing a cDNA for a FLAG-tagged human growth hormone 1 (hGH1). hGH1 is mainly synthesized as a pre-hormone in the anterior-pituitary in vivo. During processing through the endoplasmatic reticulum and the Golgi, small signaling peptides were removed, and the mature hGH1 was stored in secretory granules. GH is also synthesized in small volumes in the

References

[1] M.W. Brightman, T.S. Reese, Junctions between intimately apposed cell membranes in the vertebrate brain, J. Cell Biol. 40 (1969) 648–677.
[2] D.J. Begley, M.W. Brightman, Structural and functional aspects of the blood-brain barrier, Prog. Drug Res. 61 (2003) 39–78.
[3] D.J. Begley, ABC transporters and the blood-brain barriers, Curr. Pharm. Des. 10 (2004) 1295–1312.
[4] W. Schäbitz, C. Sommer, W. Zoder, M. Kiessling, M. Schwaninger, S. Schwab, Intravenous brain-derived neurotrophic factor reduces infarct size and counter regulates bax and bcl-2 expression after temporary focal cerebral ischemia, Stroke 31 (1997) 2212–2217.
[5] J.P. Makar, G.T. Bever, I.S. Singh, W. Royal, S.N. Sahu, T.P. Sura, S. Sultana, R.T. Sura, N. Patel, S. Dhib-Jalbut, D. Trisler, Brain-derived neurotrophic factor gene delivery in an animal model of multiple sclerosis using bone marrow stem cells as a vehicle, J. Neuroimmunol. 210 (2009) 40–51.
[6] A. Scheepens, L.S. Sirimanne, B.H. Breier, R.G. Clark, P.D. Gluckman, C.E. Williams, Growth hormone as a neuronal rescue factor during recovery from CNS injury, Neuroscience 104 (2001) 677–687.
[7] K. Gustafson, H. Hagberg, B. Bengtsson, C. Brantsing, J. Isgaard, Possible protective role of growth hormone in hypoxia-ischemia in neonatal rats, Pediatr. Res. 45 (1999) 318–323.

GENE DELIVERY

[8] M. Leist, P. Ghezzi, G. Grasso, R. Bianchi, P. Villa, M. Fratelli, C. Savino, M. Bianchi, J. Nielsen, J. Gerwien, P. Kallunki, A.K. Larsen, L. Helboe, S. Christensen, L.O. Pedersen, M. Nielsen, L. Torup, T. Sager, A. Sfacteria, S. Erbayraktar, Z. Erbayraktar, N. Gokmen, O. Yilmaz, C. Cerami-Hand, Q. Xie, T. Coleman, A. Cerami, M. Brines, Derivatives of erythropoietin that are tissue protective but not erythropoietic, Science 305 (2004) 239–242.

[9] G. Grasso, A. Sfacteria, F. Meli, V. Fodale, M. Buemi, D.G. Iacopino, Neuroprotection by erythropoietin administration after experimental traumatic brain injury, Brain Res. 1182 (2007) 99–105.

[10] D. Wu, Neuroprotection in experimental stroke with targeted neurotrophins, Neurorx 2 (2005) 120–128.

[11] B. Song, H.V. Vinters, D. Wu, W.M. Pardridge, Enhanced neuroprotective effects of basic fibroblast growth factor in regional brain ischemia after conjugation to a blood-brain delivery vector, J. Pharmacol. Exp. Ther. 301 (2002) 605–610.

[12] J. Lichota, T. Skjørringe, L.B. Thomsen, T. Moos, Macromolecular drug transport into the brain using targeted therapy, J. Neurochem. 113 (2010) 1–13.

[13] W.F. Anderson, Human gene therapy, Science 256 (1992) 808–813.

[14] J. Jo, T. Ikai, A. Okazaki, M. Yamamoto, Y. Hirano, Y. Tabata, Expression profile of plasmid DNA by spermine derivatives of pullulan with different extents of spermine introduced, J. Control. Release 118 (2007) 389–398.

[15] K. Hosseinkhani, T. Aoyama, O. Ogawa, Y. Tabata, liver targeting of plasmid DNA by pullulan conjugation based on metal coordination, J. Control. Release 83 (2002) 287–302.

[16] I. Kawakami, T. Ikai, A. Okazaki, J. Jo, M. Yamamoto, M. Imamura, A. Kanematsu, S. Yamamoto, N. Ito, O. Ogawa, Y. Tabata, Efficient gene transfer by pullulan-spermine occurs through both clathrin- and raft/caveolae-dependent mechanisms, J. Control. Release 116 (2006) 76–82.

[17] J. Jo, T. Ikai, A. Okazaki, K. Nagane, M. Yamamoto, Y. Hirano, Y. Tabata, Expression profile of plasmid DNA obtained using spermine derivatives of pullulan with different molecular weights, J. Biomater. Sci. Polym. Ed. 18 (2007) 883–899.

[18] A. Okazaki, J. Jo, Y. Tabata, A reverse transfection technology to genetically engineer adult stem cells, Tissue Eng. 13 (2007) 245–251.

[19] K. Nagane, M. Kitada, S. Wakao, M. Dezawa, Y. Tabata, Practical induction system for dopamine producing cells from bone marrow stromal cells using spermine-pullulan-mediated reverse transfection method, Tissue Eng. A 15 (2009) 1655–1665.

[20] D.K. Thakor, Y.D. Teng, Y. Tabata, Neuronal gene delivery by negatively charged pullulan-spermine/DNA anioplexes, Biomaterials 30 (2009) 1815–1826.

[21] J. Jo, A. Okazaki, K. Nagane, M. Yamamoto, Y. Tabata, Preparation of cationized polysaccharides as gene transfection carrier for bone marrow-derived mesenchymal stem cells, J. Biomater. Sci. Polym. Ed. 21 (2010) 185–204.

[22] T. Yamaoka, Y. Tabata, I. Ikada, Body distribution profile of polysaccharide after intravenous administration, Drug Deliv. 3 (1993) 75–82.

[23] L. Greiffenberg, W. Goebel, R.S. Kim, I. Weiglein, A. Bubert, F. Engelbrecht, M. Stins, M. Kuhn, Interaction of Listeria monocytogenes with human brain microvascular endothelial cells: InlB-dependent invasion, long-term intracellular growth, and spread from macrophages to endothelial cells, Infect. Immun. 66 (1998) 5260–5267.

[24] F. Gossard, F. Dihl, G. Pelletier, P.M. Dubois, G. Morel, In situ hybridization to rat brain and pituitary gland of growth hormone cDNA, Neurosci. Lett. 79 (1987) 251–256.

[25] W.A. Jefferies, M.R. Brandon, S.V. Hunt, A.F. Williams, K.C. Gatter, D.Y. Mason, Transferrin receptor on endothelium of brain capillaries, Nature 312 (1984) 162–163.

[26] S. Gosk, C. Vermehren, G. Storm, T. Moos, Targeting anti-transferrin receptor antibody (OX26) and OX26-conjugated liposomes to brain capillary endothelial cells using in situ perfusion, J. Cereb. Blood Flow Metab. 24 (2004) 1193–1204.

[27] U. Bickel, Y.S. Kang, W.M. Pardridge, In vivo demonstration of subcellular localization of anti-transferrin receptor monoclonal antibody–colloidal gold conjugate in brain capillary endothelium, J. Histochem. Cytochem. 42 (1994) 1493–1497.

[28] R.D. Broadwell, B.J. Baker-Cairns, P.M. Frieden, C. Oliver, J.C. Villegas, Transcytosis of protein through the mammalian cerebral epithelium and endothelium, Exp. Neurol. 142 (1996) 47–65.

[29] T. Moos, E.H. Morgan, Restricted transport of anti-transferrin receptor antibody (OX26) through the blood-brain barrier in the rat, J. Neurochem. 79 (2001) 119–129.

3.2   STUDY II

IN VITRO DELIVERY OF SUPERPARAMAGNETIC IRON OXIDE NANOPARTICLES THROUGH BRAIN ENDOTHELIAL CELLS

Louiza Bohn Thomsen, Thomas Linemann, Jacek Lichota, Kwang Sik Kim, Gerben Visser and Torben Moos

The manuscript has been submitted to Journal of Controlled Release

IN VITRO DELIVERY OF SUPERPARAMAGNETIC IRON OXIDE NANOPARTICLES THROUGH BRAIN ENDOTHELIAL CELLS

L.B. Thomsen[a]; T. Linemann[a]; J. Lichota[a]; K. S. Kim[b], G. M. Visser[c] and T. Moos[a*]

[a]Department of Health Science and Technology, Aalborg University, Denmark
[b]Division of Pediatric Infectious Diseases, John Hopkins University, School of Medicine, Baltimore, USA
[c]Department of Infectious Diseases & Immunology, Utrecht University, The Netherlands

* Correspondence:
Louiza Bohn Thomsen,
Section of Neurobiology, Biomedicine
Department of Health Science and Technology
Fr. Bajers Vej 3B
Aalborg University
DK-9220 Aalborg East, Denmark
Phone: + 45-99443731
E-mail: lbt@hst.aau.dk

Keywords: Blood-brain barrier, SPIOs, drug delivery, TEER, SPIO

Abbreviations

BBB – blood-brain barrier
CNS – central nervous system
BCEC - brain capillary endothelial cells
TEER – trans endothelial electric resistance
SPIO – superparamagnetic iron oxide nanoparticle
MDM – magnetic dextran microspheres

ABSTRACT

The blood-brain barrier (BBB) constitutes a physical, chemical and immunological barrier making the brain accessible to only a few percent of potential drugs intended for treatment inside the central nervous system (CNS). A new approach with the purpose of overcoming the restraints of the BBB by enabling transport of drugs, siRNA or DNA into the brain is to use superparamagnetic iron oxide nanoparticles (SPIOs) as drug-carriers. The aim of this study was to investigate the ability of fluorescent SPIOs to cross the BBB facilitated by an external magnetic force. The capability of SPIOs to penetrate the barrier was shown to be significantly higher in the presence of an external magnetic force in a static *in vitro* BBB model of the BBB. Particles added to the luminal side of the *in vitro* BBB model were found in astrocytes co-cultured in remote distance on the abluminal side, indicating that particles were transported or drawn through the barrier and either taken up by or forced into the astrocytes by the external magnetic field. The SPIOs did not negatively affect the viability of the endothelial cells as revealed by a live/dead assay and by trypan blue uptake. The magnetic force-mediated dragging of SPIOs through the BBB may denote a novel mechanism for drug delivery to the brain.

INTRODUCTION
Drug delivery to the brain has proven to be a difficult task mainly due to the presence of the blood-brain barrier (BBB) formed by tightly interconnected brain capillary endothelial cells (BCECs). The impermeability properties of the BCECs are supported by astrocytes, pericytes and neurons which together form the so-called neurovascular unit [1]. The BBB excludes most molecules from entering the central nervous system (CNS) [2] and molecules must be preferably small in size and lipophilic to enter the brain [3]. In spite of being in possession of these qualities many of the carriers however fail to deliver their cargo to the brain in an amount adequate for treatment without allowing unacceptable high off-target affection.

Many drug-carriers have been created, e.g. liposomes or polyplexes, which fulfill the demands of being lipophilic and/or at the nano-size scale. A relatively new approach in the field of drug delivery is the use of magnetic nanoparticles. Hence, magnetic nanoparticles are currently being used for various purposes such as a contrast agent for magnetic resonance imaging (MRI) [4], induction of hyperthermia for tumor therapy [5], cell labeling/cell separation [6, 7], targeted therapeutics [8, 9] and magnetofection [10].

Superparamagnetic iron oxide nanoparticles (SPIOs) is a subtype of magnetic nanoparticles which are highly magnetizable and have a core of iron oxide particles composed of magnetite (Fe_3O_4) and maghemite (γ-Fe_2O_3) [11]. The SPIOs typically have a mean diameter of 50-100nm [11], and their iron oxide core exerts low toxicity, as it is gradually degraded to Fe^{2+} and Fe^{3+} in the body and enters the pool of body iron [11]. SPIOs have been shown to induce oxidative stress in murine macrophage (J774) cells, but only in doses higher than 100µg/ml [12]. Their magnetic core can be coated with lipophilic fluorescent dyes for visual detection. Furthermore, the particles can be protected by a biocompatible polymeric shell, like dextran, polysorbate or starch, or coated by phospholipids or polyethylene glycol (PEG) to prolong their presence within their circulation due to a lower capture of the particles by the mononuclear phagocyte system [9, 13, 14, 15]. A proper coat also prevents aggregation of the particles, which they otherwise tend to due to a strong magnetic dipole to dipole attraction [13, 17]. Furthermore a protective coat enables conjugation of e.g. various proteins, DNA and drugs to the surface of the SPIOs [13, 15, 16, 17].

A major advantage of the properties of SPIOs is their ability to precisely deliver their cargo to a given target organ when drawn there to by the force of a magnetic field provided by an external or implanted magnet [13, 17]. Under the influence of the magnetic field, the SPIOs are drawn towards the magnet to concentrate near its location. Delivery of SPIOs will therefore benefit from being very local and its dosing can be minimized to reduce off-target effects [13, 17].

In this study the ability of SPIOs to function as drug carriers is investigated in an *in vitro* BBB model. The SPIOs are taken up by endothelial cell and increasingly pass the intact brain endothelial cell monolayer with the aid of an external magnet to end up in a layer of astrocytes cultured in remote distance on the "brain side" of the endothelial cells.

MATERIALS AND METHODS
Materials
Transwell membrane culture inserts and plates (Corning, Thermo Fisher Scientific), fluorescent SPIOs "nano-screenMAG-D" composed of magnetite (Chemicell, Germany), mouse-anti-ZO-1 and Alexa Fluor 488 goat-anti-mouse, live/dead cell viability assay (Invitrogen, UK), Trypan Blue stain, 4',6diamidino-2-phenyindole (DAPI) (Sigma-Aldrich, Germany), mounting media and mouse anti-cow glial fibrilary acidic protein (Dako, Denmark).

Cell cultures
Immortalized human brain microvascular endothelial cells (HBMEC) were cultured in Medium 199 with L-Glutamine and HEPES (Invitrogen) with 10 % Fetal Calf Serum (Invitrogen), 10 % Nu Serum IV (BD Biosciences, USA) and 100 U/mL Penicillin G Sodium and 100µg/mL Streptomycin sulphate (Invitrogen) [18]. Immortalized rat brain astrocytes (DI-TNC1) (ATCC, Sweden) were cultured in DMEM/F12 (Lonza, Switzerland) with 10 % fetal calf serum and 100 U/mL Penicillin G Sodium and 100µg/mL Streptomycin sulphate.

Establishment of an *in vitro* BBB model in Transwell membrane plates
HBMECs were seeded as monocultures in inserts of twelve wells Transwell membrane culture plates in a density of 150.000cells/insert. The HBMECs were cultured in an astrocyte conditioned media (ACM) consisting of a mixture of 50% DI-TNC1 media aspirated from astrocytes after 24 hours incubation and 50% HBMEC media. When mentioned DI-TNC1s were seeded in the wells of the 12 well culture plates with 100.000cells/well. The astrocytes cultured in DI-TNC1 media were grown overnight in a humidified incubator with 5% CO2 to ensure proper cell attachment. Then the inserts containing HBMECs were re-inserted into the Transwell culture plate's containing the DI-TNC1 astrocytes to form a non-contact co-culture. The medium was replaced every day to avoid high media changes in the pH.

Trans Endothelial Electrical Resistance (TEER) measurements
TEER measurements were conducted with a Millicell™ ERS-2 apparatus (Millipore, USA) and an STX-1 electrode (Millipore). To calculate the TEER R_{blank} was subtracted from R_{Sample}, and then the product was multiplied by the well area. The TW in this study had a well area of $1,1 cm^2$; therefore the equation was as follows: $(R_{Sample} - R_{blank}) \times 1.1\ cm^2 = \Omega^* cm^2$

The TEER was measured every second day up until seven days, and thereafter every day. Just before the TEER measurements were made the culture media was changed and cells and media were allowed to reach room temperature. Three measurements were made on each well from which an average TEER value was calculated.

Fluorescent SPIOs
The SPIOs used in this study is commercially available magnetic iron oxide nanoparticles with a hydrodynamic diameter of 100nm. They consist of a magnetic magnetite core surrounded by a lipophilic fluorescence dye covered by a polysaccharide matrix of starch consisting of α-D-glucose units (Fig. 1). Both red and blue fluorescent SPIOs were used in this study. The blue fluorescent nanoparticles have maximal excitation at 378 nm and emission at 413 nm and red nanoparticles have excitation wavelength at 578 nm and emission wavelength at 613 nm.

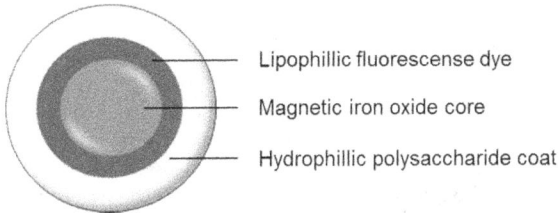

Lipophillic fluorescense dye

Magnetic iron oxide core

Hydrophillic polysaccharide coat

Figure 3-1 The SPIOs consist of a magnetic core, surrounded by a lipophilic fluorescence dye layer covered by a polysaccharide matrix of starch consisting of α-D-glucose units.

Size and charge of the SPIOs
Size (DLS/Non-Invasive Back-Scatter (NIBS)) equivalent to particle diameter and charge/ ζ- potential were measured on a Zetasizer Nano (Malvern, UK). 20 µg of SPIOs was diluted in 1 ml double distilled water and tested in triplicate. The size of the SPIOs was analyzed based on the Culmulants method by the computer software which calculated the Rs values and provided the apparent size of the SPIOs. The ζ-potential was likewise calculated by the software tested tree times.

Application of SPIOs on the BBB model
When the TEER of the HBMEC's reached a plateau, indicating that the highest TEER had been reached and the endothelial cells had formed a barrier, the fluorescent SPIOs (Fig.1) were added to the inserts in doses of 35, 70 and 140

µg/insert in three replicas of each concentration. The process of addition of SPIOs to the cell culture is demonstrated in Fig. 2.

The external magnetic force was supplied by a ferrite block magnet with field strength of 0.39 Tesla.

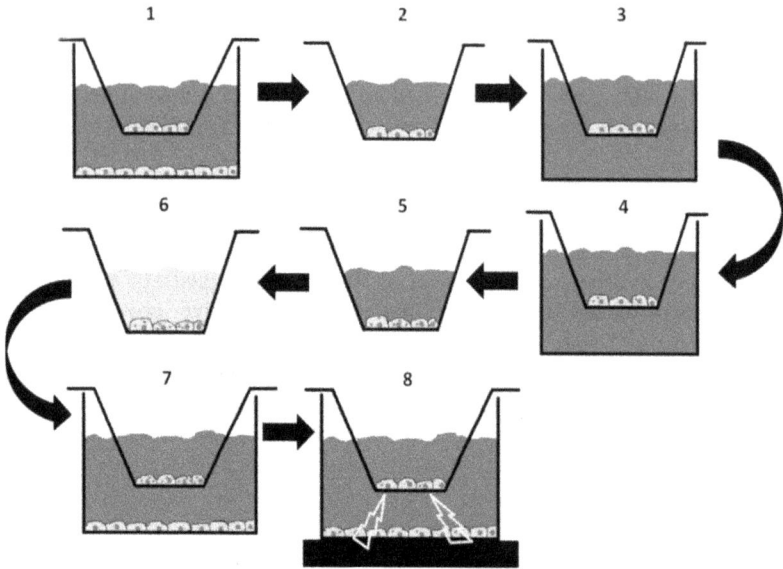

Figure 3-2 Drawing of the method employed for addition of SPIOs to the culture inserts with the aid of avoiding paracellular transport. SPIOs are depicted in blue. 1) The inserts were moved (2) to other twelve well plates (3). After addition of the nanoparticles (4), the endothelial cells (pink) were incubated for 24 hours. Afterwards, the media from the inserts was changed, and the inserts washed three times with PBS to remove nanoparticles that had not been taken up by the HBMEC's (5+6). The inserts containing endothelial cells in PBS (light blue) (6) were reinserted containing cells in the twelve well plates (7) and placed on a ferrite block magnet for 5 hours at 37°C to draw the magnetic particles towards the bottom of the well (8) in where astrocytes (green) were cultured in remote distance from the endothelial cells. A control plate was also kept at 37°C in secure distance from the magnet. After 5 hours the media in upper and lower chambers of the wells was collected and stored at 4°C. A control plate was also kept at 37°C in secure distance from the magnet. After 5 hours the media of the upper and lower chambers of the wells were collected and stored at 4°C.

Immunostaining

After terminating the experiment, the HBMEC's of the control and the experimental plates were washed three times in PBS, fixed in 4% paraformaldehyde for 4 minutes and washed three times in PBS. The cells were incubated overnight with mouse-anti-zonula occludens-1(ZO-1), and binding of the primary antibody was visualized using Alexa Flour 488 goat-anti-mouse. DI-TNC1s were incubated overnight with mouse-anti-glial-fibrillary acidic protein (GFAP), and binding of the primary antibody was visualized using Alexa Flour

488 goat-anti-mouse. The cell nuclei of both DI-TNC1s and HBMECs were stained with DAPI for 5 minutes. The membrane of the inserts containing HBMECs was cut out of the insert, mounted on a slide with fluorescent mouting media and observed under a fluorescence microscope.

Cytotoxicity

To examine if the cells gets impaired by the SPIOs or by the application of the external magnetic field, the cell viability was visualized using a live/dead cell viability assay. The assay was performed according to the recommendations from the vendor. In brief, two working solutions were prepared: Solution one containing 50 µM C12-resaurin in Dimethylsulfoxide (DMSO), solution two consisting of 1 M SYTOX Green stain in DMSO. The culture medium was aspirated from the TW inserts, and 0.25 ml of PBS added to each well. The working solutions were added to the wells to reach a final concentration of 5 µM C12-resazurin and 50 nM SYTOX Green dye in the two solutions respectively. The cells where then incubated at 37°C in an atmosphere of 5% CO_2 for 15 minutes and afterwards they were kept on ice, rinsed three times with PBS and observed under a fluorescence microscope.

   Dead and viable cells were counted on the basis of a counting of Trypan blue-labeling, as trypan blue only enters dead cells. Cells were cultured in monoculture in six wells culture plates until 100 % confluence was reached. Then SPIOs was added to half of the wells in a concentration of 1170 µg which corresponded to the highest dose (140µg/insert) added in amount per square centimeter in the experiment described above. The cells were incubated with or without SPIOs for 24 hours and placed on the plate magnet for 5 hours. The cells were then trypsinized and mixed with Trypan blue. An appropriate amount of cell suspension containing Trypan blue was then filled in a hemocytometer and dead and living cells counted. The total amount of dead and vital cells were calculated and a student's T-test was performed to test, if there were any differences in the amount of vital and dead cells between the control wells and experimental wells subjected to the magnetic force. A p-value at $p < 0.05$ was considered statistically significant.

Quantification of SPIOs crossing the BBB *in vitro*

The well plates with the presence of DI-TNC1 astrocytes were investigated under a fluorescence microscope with the medium remaining in the wells. The fluorescent SPIOs were counted using a counting mesh with an area of 0,054mm^2 that was inserted inside of the microscopes ocular. Counts were made at randomly picked areas 10 times per well to obtain a statistical correct counted average of the amount of nanoparticles in the wells. By the use of a student's t-test, it was examined if there were any differences between the amounts of particles in the wells of control versus experimental plates. A p-value was considered statistically significant at $p < 0.05$.

RESULTS AND DISCUSSION

In this study we wanted to investigate whether SPIOs are able to enter and cross BCECs and if an external magnetic force could be applied to aid the penetration rate and efficiency. We also wanted to test if the particles have a toxic effect on the cells and will obstruct the barrier when passing the endothelial cell layer.

Size and charge of the SPIOs

The hydrodynamic diameter of the SPIOs was determined by DLS, which is a back scatter analysis. The SPIOs had a mean diameter of 117.5 nm which is a little larger than proclaimed by the manufacturer. Furthermore the ζ-potential of the SPIOs was measured to be -16.8 mV. Starch coated SPIOs have previously been found to be of similar anionic charge [19].

SPIOs enter into and cross though endothelial cells

Using immunofluorescence, the HBMECs were investigated for their expression of ZO-1, a marker of tight junctions, before and after exposure to SPIOs and subsequent magnetic force (Fig. 3). Clear signal and equal intensity of the ZO-1 marker protein provide morphological evidence that tight junctions were present between HBMECs in both the experimental and control plates.

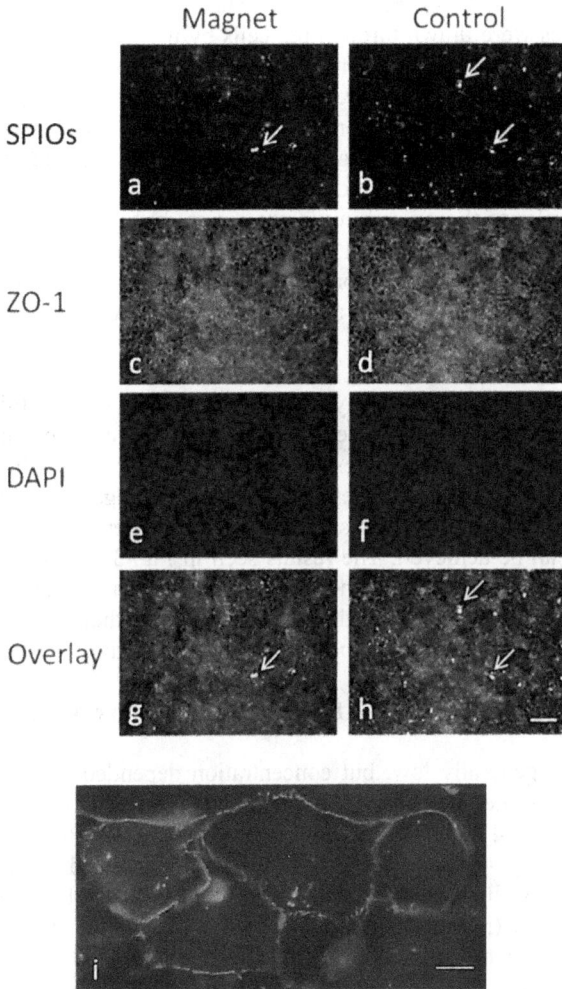

Figure 3. The presence of fluorescent SPIOs (a,b), ZO-1 (c,d), and DAPI (e,f) in inserts with cultured HBMECs in absence (b, d, f, h) or presence (a, c, e, g) of magnetic force. Overlay of a, c, e in g and overlay of b, d, f in h. A magnification of h can be seen in i. The ZO-1 expression is prominent in both the experimental and control plates (a and b). Notice that there are fewer SPIOs present in the insert exposed to a magnetic force (a) compared to that of the control insert (b). (The red fluorescent SPIOs are shown in white for better visualization) (The insert membranes tend to bulge on the slides which results in different levels of focal points within a frame and "out of focus areas" can be observed) ((a-h) scale bar = 50μm, (i) scale barr = 5μm)

Before applying the external magnetic force, SPIOs were located inside the HBMEC monolayer (Fig. 3). After exposition to the magnetic force fewer particles were detected in the HBMEC monolayer (compare Figs. 3a and 3b), indicating that the nanoparticles were drawn through the cells by the magnetic force and entered the abluminal ("brain side") chamber of the microporous membrane. These findings suggest that the SPIOs are taken up by the HBMECs even without the addition of an external magnetic force. They also indicate that SPIOs do not need any further chemical/physical changes on their surface to interact with the BCECs and can subsequently be internalized by BCECs.

SPIOs very similar to those used in the present study coated with starch and of a size of ~110nm have been shown to enter the brain of Fischer 344 rats when injected intravenously without the presence of a magnetic force [13]. This supports the findings in the present study and it seems as there are to some extent an extravasation of the SPIOs.

These results imply that a targeting strategy towards BCECs is needed if these are to be the only target. If systemically injected the SPIOs used in this study would probably also interact with other cells than BCECs . Therefore a new strategy is necessary for targeting the particles to the BCECs only. It has been shown that SPIOs can be coated with substrates that can bind e.g. ligands or antibodies [15]. With such modified SPIOs, BCECs can be directly targeted and exclusive uptake by in BCECs can be achieved. The results seen in Fig. 3 also suggest that SPIOs can be drawn out of the HBMEC monolayer by an external magnetic force, which contributes the targetability aiding their passage towards their intended destination. This phenomenon was therefore explored further in this study.

Exposure to SPIOs and magnetic force does not lead to cytotoxicity of the endothelial cells.

SPIOs exhibit a generally low, but concentration depended cytotoxic action [11, 12]. Our study revealed no signs of lost vitality of the HBMECs after the cells had been incubated for 24 hours with various concentrations of SPIOs (35µg, 70µg and 140 µg per ml) (Fig. 4). Hence, a trypan blue stain conducted to count the amount of dead cells in wells incubated with or without 140µg/ml SPIOs revealed no statistical difference (p<0,05) between cell viability in the two conditions. Naqvi et al. (2010) observed the toxicity of SPIOs with a Tween 80 coat and 30nm in diameter increases in a concentration-dependent manner [12]. In their measurements, the toxicity seen as a marked change in cell viability was observed when between 100 and 200µg/ml SPIOs were added to cultures of murine macrophage cells (J774), indicating that SPIOs are non-toxic to cells in concentrations of 100µg/ml or less [12]. These data are in good accordance with the results of the present study even though the concentration of 140µg/ml lies within their range of a toxic concentration, but does not exhibit any toxic effect on cells in our study. Furthermore it has been shown that SPIOs with an anhydroglucose polymer coat and 50-150nm in diameter did not affect the mortality of Sprague-Dawley rats when injected in the tail vein in a dose of 5% of the estimated blood volume [20]. The rats were monitored for up to 65 days and it was detected that the amount of magnetic particles found in the animal decreased over time [20]. These data indicates that magnetic nanoparticles can be administered systemically without exerting toxicity on cell cultures or animals. It also seems that when given *in vivo* the particles are cleared probably by deposition into iron stores in the cells. Therefore the particles should be safe to administer in small doses *in vivo*.

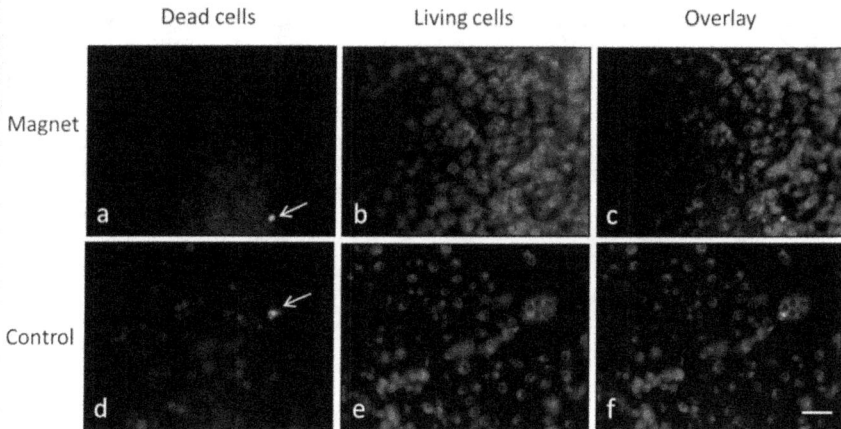

Figure 4. Live/dead cell viability stain of HBMEC's in Transwell membrane inserts. Lower row, HBMEC's in inserts from experimental plates subjected to magnetic force. Upper row, HBMEC's in inserts from controle plates. 70 µg of SPIOs were added to the insert on both plates. Dead cells are visualized with Sytox green stain/uptake (a+d)(indicated by white arrows). Live cells are visualized with resazurin staining (b+e). Overlays of the two stains are seen on c) and f). There are no differences in the viability between the cells of the two plates. (scale bar = 50µm)

The integrity of the in vitro BBB model

A commonly used *in vitro* model of the BBB consisting of a microporous membrane insert to form a static model of the BBB was employed in this study. In Figure 5 the TEER values measured on HBMECs of a control and experimental plate can be seen. The TEER values was rather low but did reach a threshold TEER value of 43.6 ± 0.7 $\Omega{*}cm^2 \pm SE$ which is not unusual for human brain microvascular cell lines [21]. The TEER depends on how tightly the BCECs are interconnected via tight junctions and a low TEER could indicate that there are open areas in between BCECs [21]. Although the immunostaining of ZO-1 showed the presence of tight junctions between BCECs we wanted to eliminate the possibility of SPIOs passing the BCECs paracellularly.

To secure exclusive transcellular transport of SPIOs by the HBMECs and exclude paracellular transport, the cells were cultured in culture inserts and removed from the culture plates (empty or containg astrocytes) while adding the SPIOs (see description in the results section and Fig. 2). After incubation with the SPIOs for 24 hours the HBMEC monolayer was washed to remove excess nanoparticles and the inserts were returned to their original culture plate (Fig. 2). By performing this step it was ensured that any particles detected in the abluminal side of the well-chamber under the inserts could only derive from the magnetic force drawing the particles through the HBMECs or from their secretion.

The TEER of the HBMECs reached a plateau after approximately 6 days of culture (Fig. 5). The TEER values were also measured after magnetic force had been applied to the HBMECs containing SPIOs (Fig. 5). The stable TEER values after the exposure to the magnetic field indicate that the integrity of this *in vitro* BBB was not harmed by the magnetic-field-aided penetration of the SPIOs. This observation is in good agreement with the findings in Saiyed et al. (2010) who showed that magnetic particles encapsulated in liposomes were taken up by monocytes and drawn through an *in vitro* BBB model with an external magnet without affecting TEER values [17].

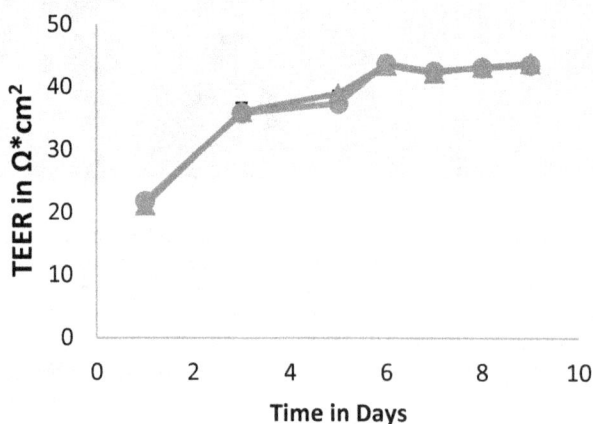

Figure 5 The graphs shows the measured TEER values over time from both the control (circle) and experimental plate (triangle) in HBMEC monocultures grown in the presence of a astrocyte-conditioned media. The fluorescent SPIOs were added at experimental day 8, and at day 9 the cell culture plate was placed on a block magnet for five hours and TEER measured afterwards. The TEER values of the epithelial monolayer peaked at day 6 and did not decrease after the passage of nanoparticles. Hence, the TEER values of both curves are stabile before and after the application of an external magnetic force and indicates that the barrier had not been obstructed by the passage of the particles through the endothelial cells (n = 11, results presented as means ± standard error (SE) (very low SE values)).

Passage of SPIOs through the BBB *in vitro*

The SPIOs crossed the HBMEC monolayer under the influence of an external magnetic field (0,39T), and their passage occurred in a concentration dependent manner (Fig. 6a-c). This indicates that SPIOs can be drawn through the BBB and into the brain parenchyma. A limited number of SPIOs were observed in the lower chamber without exposure to the magnetic field (Fig. 6d-f). However, this number was very low and did not seem to increase when increasing the concentration of the nanoparticles. This suggests that the SPIOs are able to pass the *in vitro* BBB in a low concentration without the external magnetic force.

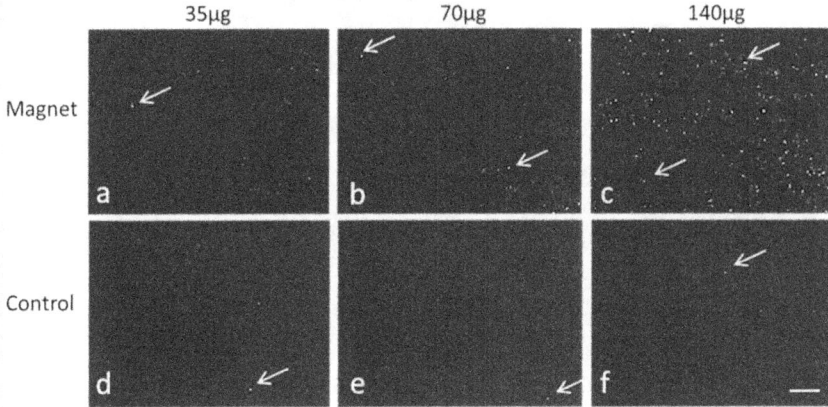

Figure 6 SPIOs in monocultures of HBMECs cultured in cell culture inserts. For better visualization the red fluorescent SPIOs are here shown in white. Upper row: Cells with exposure to the magnet for 5 hrs. Lower row: Cells without exposure to the magnet. The pictures shows the presence of fluorescent nanoparticles in wells of the lower chamber under the inserts to which 35 μg (a+d), 70 μg (b+e), or 140 μg (c+f) SPIOs have been added. Arrows indicate some of the magnetic particles. The concentration of nanoparticles is visually higher in the wells of the experimental plates than in the wells of the control plates. (Scale bar = 30μm)

The amount of particles that penetrated the BCECs and entered into the astrocytes cultured on the bottom of the wells was additionally quantified in the control and experimental plates (Fig. 7). The experimental plate was submitted to an external magnetic field for 5 hrs. 35, 70 or 140 μg of SPIOs were added to inserts containing BCECs. The result of the quantification of particles entering the astrocytes under the inserts can be seen in figure 8.

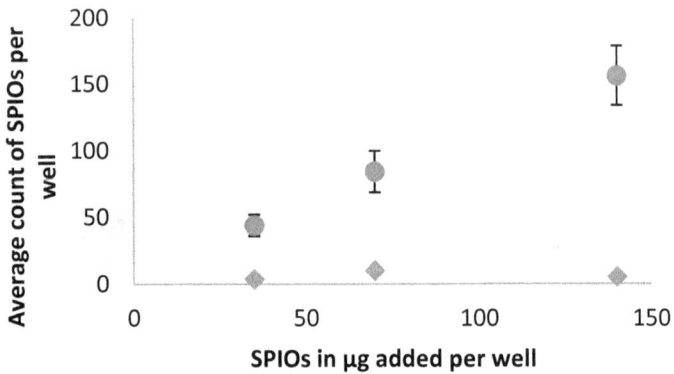

Figure 7. The graph depicts the relation between the amount of magnetic nanoparticle passing through BCECs and into DI-TNC1 cultured on the bottom of the wells and the concentration of SPIOs added to the BCECs. Experimental plate (circle) which was submitted to an external electric field for 5 hrs and control plate (square) devoid of an external electric field subjection. SPIOs were added to BCECs in concentrations of 35, 70 or 140µg per insert. The amount of SPIOs passing the BCECs an entering into astrocytes was clearly higher when an external magnetic field was applied. There seems to be a linear correlation between dose and response when applying the magnetic field (n= 4 observations per point at 70 and 140 µg and n=3 observations per point at 35µg, results presented as means ±SE).

There is a statistically significant difference between the counts from wells of the experimental plate and the control plate (35µg: p<0,001, 70µg: p<0,001, 140µg: p<0,001). The passage of SPIOs across BCECs was increased 11, 8 and 29-fold over the control at a concentration of 35, 70 and 140µg/ml respectively. Chertok et al (2008) observed that a magnetic force of 0.4T increased the concentration of starch coated SPIOs (~110nm) targeted towards a rat brain tumor by 11.5-fold over the amount found in non-targeted (no magnetic force applied) brain tumors [13]. Similar results have been shown for starch coated SPIOs with a diameter of 46nm which was intravenously injected into nude mice with armpit tumor xenografts [22, 23]. The SPIOs were shown to accumulate in a higher concentration in the tumors when subjected to an external magnetic field of 0.5 T [22, 23]. These studies all refer to magnetic-force-increased delivery of SPIOs in tumor tissue which is known to have a compromised blood-tumor barrier. Chertok et al (2008) found that the concentration of SPIOs dispersed into normal brain tissue of Fischer 344 rats seemed to increase slightly (approximately 3-fold) under the influence of a magnetic field (0.4T) over non-magnetic-force-targeted SPIOs [13]. Also the migration of monocytes loaded with magnetic liposomes has been shown to be enhanced 3-fold by applying a magnetic force in an *in vitro* BBB model [17]. In the present study passage of SPIOs across an *in vitro* BBB in non-toxic doses was clearly increased by the magnetic field. The rate of SPIOs penetrating the BCECs without any aid of an external magnetic field was low and did not significantly change in spite of a change in dose concentration. In the experimental plate the concentration of SPIOs added to the inserts containing BCECs linearly correlated with the concentration entering the astrocytes cultured in wells when submitted to an external magnetic field. Hence, these results support the strategy of employing SPIOs for targeted delivery to the brain.

SPIOs pass through the in vitro BBB and further into astrocytes
Next the ability of the SPIOs to cross the in vitro BBB and to enter cells on the "brain side" was investigated. To answer these questions astrocytes were cultured on the bottom of the wells but otherwise the experimental setup was the same as in the previous section with HBMECs cultured in inserts inserted in the wells containing astrocytes. Fluorescent SPIOs were found in the astrocytes in both the experimental plates (Fig. 8) and in controls.

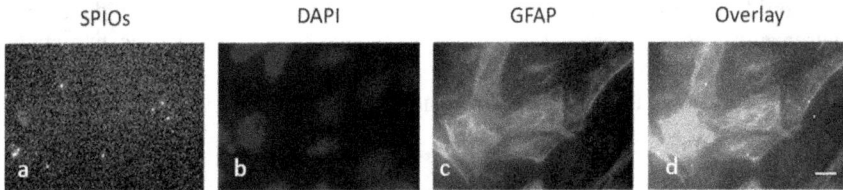

Figure 8. Fluorescence microscopy showing SPIOs (a) inside DAPI (b) and GFAP (c) stained astrocytes present at the bottom of a lower well of a culture plate. Overlay of a,b, and c is seen in d. The astrocytes are from an experimental plate that has been exposed to a magnetic field for 5 hrs. The pictures show the presence of fluorescent SPIOs inside astrocytes cultured in a plate well under an insert to which 70 μg SPIOs was added. The red fluorescent SPIOs are shown in white for better visualization. (Scale bar = 15μm)

Differences were observed with respect to the uptake of SPIOs in astrocytes as only a minimal amount of particles was observed in astrocytes of the control plates as compared to that of astrocytes of the experimental plates. This suggests that the SPIOs can not only be drawn through the *in vitro* BBB but also enter cells present in remote distance of the abluminal side of the barrier. As earlier mentioned, Chertok et al (2008) observed presence of intravenously injected SPIOs in the brain parenchyma of normal rat brain tissue [13]. Hence, a possible application of the SPIOs for drug-delivery not only applies to BCECs but also to neurons and glial cells located deeper inside the brain. This notion, together with the fact that the magnetic particles are capable of movement in a particular direction via the application of an external magnetic field, signifies these magnetic particles as potential drug-carriers. The SPIOs are therefore obvious candidates as drug-carriers for CNS drug-delivery beyond the BBB, and their specific uptake by BCECs may be improved by conjugation to a targeting molecule.

CONCLUSIONS

The SPIOs could pass into and through the BCEC monolayer and enter astrocytes cultured at the bottom of the lower chambers in a manner that was clearly enhanced by the use of an external magnetic force. The external magnetic force did not affect the integrity of the endothelial monolayer, neither was the cell viability affected by the fluorescent SPIOs or by the magnetic force dragging the nanoparticles through the cells.

Our main conclusion is therefore that SPIOs can be used for penetration of the BCECs and further into the brain without harming the cells. SPIOs can be conjugated with various compounds and our results are indicative of SPIOs as nano-carriers for future drug-delivery (purpose involving targeted therapeutics) to the brain.

ACKNOWLEDGEMENTS

We would like to thank Joachim Høeg Mortensen, Christian Garn du Jardin Nielsen and Thomas Larsen for their input to this study. The data in this study were generated by kind grant support from the Danish Medical Research Council (grant no. 271-06-0211), the Spar Nord Fund, and the Obelske Family Fund.

REFERENCES

[1] F.L. Cardoso, D. Brites, M.A. Brito, Looking at the blood-brain barrier: Molecular anatomy and possible investigation approaches, Brain Res Rev, 64 (2010) 328-63.

[2] M.W. Brightman, T.S. Reese, Junctions between intimately apposed cell membranes in the vertebrate brain, J. Cell Biol., 40 (1969) 648-677.

[3] Lichota, J.; Skjørringe, T.; Thomsen, L.B.; Moos, T. Macromolecular drug transport into brain using targeted therapy. *J. Neurochem*, 113 (2010) 1-13.

[4] Bulte, J.W.; Kraitchman, D.L. Monitoring cell therapy using iron oxide MR contrast agents. Curr Pharm Biothecnol., 6 (2004) 567-584

[5] Rodríguez-Luccioni, H.L.; Latorre-Esteves, M.; Méndez-Vega, J.; Soto, O.; Rodríguez, A.R.; Rinaldi, C.; Torres-Lugo, M. Enhanced reduction in cell viability by hyperthermia induced by SPIOs, Int J Nanomedicine, 6 (2011) 373-380

[6] J. Ruan, J. Shen, Z. Wang, J. Ji, H. Song, K. Wang, B. Liu, J. Li, D. Cui, Efficient preparation and labeling of human induced pluripotent stem cells by nanotechnology, Int J Nanomedicine, 6 (2011) 425-435

[7] R. Gordon, C.E. Hogan, M.L. Neal, V. Anatharam, A.G. Kanthsamy, A. Kanthasamy, A simple magnetic separation method for high-yield isolation of pure primary microglia, J Neurosci methods, 194 (2011) 287-296

[8] Jain, T.K.; Richey, J.; Strand, M.; Leslie-Pelecky, D.L.; Flask, C.A.; Labhasetwas, V. SPIOs with dual functional properties; drug delivery and magnetic resonance imaging. Biomaterials, 29 (2008) 4012-4021

[9] Yallapu, M.M.; Foy, S.P.; Jain, T.K; Labhasetwar, V. PEG-functionalized SPIOs for drug delivery and magnetic resonance imaging applications, Pharm Res, 27 (2010) 2283-2295

[10] Scherer, F.; Anton, M.; Schillinger, U.; Henkel, J.; Bergeman, C.; Kruger, A.; Gänsbacher, B.; Plank, C. Magnetofection: enhancing and targeting gene delivery by magnetic force in vitro and in vivo. Gene Therapy, 9 (2002) 102-109
[11] Boyer, C.; Whittaker, M.R.; Bulmus, V.; Liu, J.; Davis, T.P. The design and utility of polymer-stabilized iron oxide nanoparticles for nanomedicine applications, NPG Asia Mater, 2 (2010) 23-30.
[12] Naqvi, S.; Samim, M.; Abdin, M.Z.; Ahmed, F.J.; Maitra, A.N.; Prashant, C.K.; Dinda, A.K. Concentration-dependent toxicity of iron oxide nanoparticles mediated by increased oxidative stress, Int J Nanomedicine, 5, (2010) 983-989
[13] Chertok, B.; Moffat, B.A.; David, A.E.; Yu, F.; Bregemann, C.; Ross, B.D.; Yang, V.C. Iron oxide nanoparticles as a drug delivery vehicle for MRI monitored magnetic targeting of brain tumours, *Biomaterials*, 29 (2008) 487-96.
[14] Kumar. A;, Jena, P.; Behera, S.; Lockey, R.; Mohapatra, S.; Mohapatra, S. Multifunctional SPIOs for targeted delivery, Nanomedicine, 6 (2009) 64-69.
[15] Denizot, B.; Tanguy, G.; Hindre, F.; Rump, E.; Jeune, J.; Jallet, P. Phosphorylcholine coating of iron oxide nanoparticles, J Coll Interf Sci, 209 (1999) 66-71.
[15] Ku, S.; Yan, F.; Wang, Y.; Sun, Y.; Yang, N.; Ye, L. The blood-brain barrier penetration and distribution of PEGylated fluorescein-doped magnetic silica nanoparticles in rat brain, Biochem Biophys Res Commun., 394 (2010) 871-876
[16] Dias, A.M.G.C.; Hussain, A.; Marcos, A.S.; Roque, A.C.A.   A biotechnological perspective on the application of iron oxide magnetic colloids modified with polysaccharides, Biotech adv, 29 (2011) 142-155.
[17] Saiyed, Z.M.; Gandhi, N.H.; Nair, M.P.N. Magnetic nanoformulation of azidothymidine 5'-triphosphate for targeted delivery across the blood-brain barrier, *Int J Nanomedicine*, 5, (2010) 157-66.
[18] Thomsen, L.B.; Lichota, J.; Kim, K.S.; Moos, T. Gene delivery by pullulan derivatives in brain capillary endothelial cells for protein secretion, J Control Release, 151 (2011) 45-50.
[19] Kim, D.K.; Mikhaylova, M.; Wang, F.H.; Kehr, J.; Bjelke, B.; Zhang, Y.; Tsakalakos, T.; Muhammed, M. Starch-coated superparamagnetic nanoparticles as MR contrast agents, Chem Mater, 15 (2003) 4343-4351
[20] Lubbe, A.S.; Bergemann, C.; Huhnt, W.; Fricke, T.; Riess, H.; Brock, J.W.; Huhn, D. Preclinical experiences with magnetic drug targeting: tolerance and efficacy, Cancer Res, 56 (1996) 4694-4701
[21] Deli, M.A.; Abraham, C.S.; Kataoka, Y.; Niwa, M. Permeability studies on in vitro blood.brain braiier models: physiology, pathology, and pharmacology, Cell Mol Neurobiol, 25 (2005) 59-127
[22] Jiang, J.S.; Gan, Z.F.; Yang, Y.; Du, B.; Qian, M.; Zhang, P. A novel magnetic fluid based on starch-coated magnetite nanoparticles functionalized with homing peptide, J Nanopart Res, 11 (2009) 1321-1330[23] Yang, Y.; Jiang, J.S.; Du, B.; Gan, Z.F.; Qian, M.; Zhang, P. Preparation and properties of a novel drug delivery system with both magnetic and biomolecular targeting, J Mater Sci: Mater Med, 20 (2009) 301-307

3.3   STUDY III

DYNAMIC VERSUS STATIC IN VITRO BLOOD-BRAIN BARRIER MODELS

Louiza Bohn Thomsen and Torben Moos

The manuscript displays unpublished results

Dynamic versus Static *in vitro* Blood-Brain Barrier Models

Louiza Bohn Thomsen* and Torben Moos

Department of Health Science and Technology, Biomedicine, University of

Aalborg, Denmark

*Correspondence:
Louiza Bohn Thomsen
Department of Health Science and Technology,
Biomedicine, University of Aalborg
Frederik Bajers Vej 3B,
9220 Aalborg East, Denmark
E-mail: lbt@hst.aau.dk

Keywords: Blood-brain barrier, DIV-BBB, in vitro BBB modeling, static BBB model, TEER

Abstract

The Blood-Brain Barrier (BBB) is a functional barrier preventing passage of certain compounds from the blood to the brain. When addressing complex issues, such as drug-delivery to the brain, it is important to understand the physiology of the BBB. Different models have been developed to mimic the BBB for this purpose. Applying an *in vitro* BBB model is a more ethical and less expensive method. Recently a new and improved dynamic *in vitro* BBB model (DIV-BBB) was developed by Flocel Inc. This model should be able to mimic the natural state physiological permeability properties of the BBB. This is not possible to mimic in static *in vitro* BBB models with hanging culture inserts. In the new DIV-BBB, unlike the static BBB models, cells can be grown in hollow fibers mimicking blood vessels and exposed to a pulsating flow of media mimicking the blood flow. The flow induces shear stress and this factor has shown to be of great importance when forming a tighter BBB. In this study the static *in vitro* BBB model and the DIV-BBB model are tested individually and compared afterwards. The static in vitro BBB model produced the tightest BBB when BCECs was cultured in a contact co-culture with astrocytes with 550nM hydrocortisone added to the culture media. It was not possible to produce any reliable results with the dynamic in vitro BBB in this study. The static in vitro BBB model therefore proved to be the most reliable model.

Abbreviations:
BBB: Blood-brain barrier
DIV-BBB: Dynamic *in vitro* blood-brain barrier
TEER: Trans endothelial electrical resistance

Introduction

The Blood-Brain Barrier (BBB) is formed by specialized brain capillary endothelial cells (BCECs), which form the walls of the blood vessels in the brain [1]. The BCECs are surrounded by a basal membrane which they form together with adjacent astrocytes. Astrocytes contact and cover most of the abluminal side of the BCECs with their end-feet [2, 3]. Pericytes also make contact with the BCECs and are found in the basal membrane in between the astrocytes and BCECs [4, 5]. Furthermore neurons have been found to make contact with BCECs [6].

BCECs make intercellular contacts called tight-junctions. Tight-junctions prevent leakage of substances into the brain, by preventing the passage of substances in between the BCECs [7, 8]. As a result substances can only enter the brain at the BBB through the BCECs either by diffusion or by transport via carrier-mediated transporters [3, 9, 10]. In this way diffusion/transport across the barrier can be strictly modulated by the BCECs [11]. The formation of tight-junctions and other features of the BBB characteristics are induced and maintained by astrocytes, pericytes and possibly also neurons [4, 6, 12, 13, 14, 15].

It has been demonstrated that shear stress, generated by the flow of blood across the endothelial cell surface, is an important factor in regulation of the genetic and physiological properties of the BBB [16]. The tightness of the BBB increases when BCECs are exposed to flow [9, 17, 18, 19]. Also up regulation of cAMP by hydrocortisone addition to the BCECs have shown to strengthen the BBB properties and thereby heighten the barrier integrity [10, 20].

The tightness of the BBB can e.g. be measured by recording the trans-endothelial electrical resistance (TEER). TEER is the electrical resistance formed across the BCECs and provides a measure of the barrier integrity. The tighter the barrier is the higher TEER values can be measured because the passage of electrons across the cells decreases and therefore creates a difference in the electric potential [11].

Administering e.g. drugs to the brain have been shown to be difficult because of the BBB properties just described. The solution to the problem could be drug-delivery, where drugs are carried into the brain over the BBB by a drug-carrier. For testing such drug-carriers abilities to penetrate the BBB an experimental setup is needed. It has been difficult to make direct observations of the BBB physiology on living animals, and therefore different *in vitro* models with cultured cells have been developed for this purpose [11]. Hanging cell culture inserts are static *in vitro* BBB models, which have been applied for many years. In these models the BCECs can be cultured alone or co-cultured with astrocytes, pericytes or neurons or in a combination of one or more of the cell types [5, 20, 21,

22, 23, 24]. The BCECs will form a BBB in these inserts, but the TEER values measured in this model does not always mimic the TEER values (between1200-8000 Ω*cm^2) measured in living animals and does not induction of shear stress [9, 25]. In a new dynamic *in vitro* blood-brain barrier (DIV-BBB) model developed by Flocel Inc. the BCECs are grown inside hollow fiber tubes, that mimics blood vessels, and astrocytes are grown on the outside of the fibers, supporting the BCECs. The fibers are placed in a sealed chamber, where they are exposed to a pulsatile flow, which passes through the fibers, mimicking the blood flow through the vessels. The TEER was measured by the manufacturer to be ~1200 Ω*cm2 in the DIV-BBB model and therefore mimic the in vivo BBB more closely than most static models [18, 19, 26, 27, 28, 29]. In this study a static and a dynamic *in vitro* BBB model are tested individually and compared based on the tightness of the barriers measured in TEER.

Materials and methods

Cell culture
Three kinds of immortalized endothelial cell cultures were used in this study. Human Brain Microvascular BCECs (HBMEC) were kindly provided by Professor Kwang Sik Kim, Johns Hopkins Univ. School of Medicine, Baltimore. HBMEC's were grown in culture flasks precoated with collagen (5µg/ml, BD Biosciences) in growth medium consisting of Medium 199 (Invitrogen), 10 % fetal bovine serum (Invitrogen), 10 % NuSerum (BD Biosciences) and 100 U of Penicillin G sodium per ml and 100µg Streptomycin sulfate per ml (Invitrogen). When mentioned 550nM/ml hydrocortisone (Sigma-Aldrich, Germany) was added to the HBMECs media to induce a greater tightness hence higher TEER of the HBMECs. Rat Brain BCECs (RBE4) was cultured in Alpha minimum essential medium with glutamax-1 (Gibco, Invitrogen) and Ham's F10 (Gibco, Invitrogen) in a 1:1 relation with 10% fetal calf serum (Invitrogen), 1ng/ml human basic fibroblast growth factor (Invitrogen) and 100 U of Penicillin G sodium per ml and 100µg Streptomycin sulfate per ml (Invitrogen). Mouse BCECs (Bend3) were cultured in DMEM 1885 (Sigma-Aldrich) with 10% fetal calf serum (Invitrogen), 1ng/ml human basic fibroblast growth factor (Invitrogen) and 100 U of Penicillin G sodium per ml and 100µg Streptomycin sulfate per ml (Invitrogen).

The astrocytes used in this study were either Human Astrocytes (HAs) (Sciencell cat no 1800) or rat brain astrocytes DI-TNC1 (ATCC). HAs were grown in culture flasks precoated with Poly-L-Lysine (3µg/ml, Sigma-Aldrich). Both HA and DI-TNC1 were grown in DMEM-F12 (Biochom AG), 5-10 % fetal bovine serum (Invitrogen) and 100 U of Penicillin G sodium per ml and 100µg Streptomycin sulfate per ml (Invitrogen).

The Static in vitro BBB models setup
BCECs and astrocytes were cultered in either 12-well Transwell-Clear Polyester Membrane plates (Costar) with hanging cell culture inserts (RBE4 and Bend3) or

in 12-well culture plates (Costar) with hanging Millicell culture inserts (HBMEC) in each well (Figure 1). The experiments were performed in four setups. 1) Either the BCECs were cultured alone, 2) in non-contact co-culture with astrocytes, 3) in contact co-culture with astrocytes or 4) in contact co-culture with astrocytes with media containing 550nM hydrocortisone. In contact co-culture the astrocytes $(1,0x10^5$ cells/insert) were seeded under the bottom of the inserts and in non-contact co-culture in the wells of the culture plates $(1,5x10^5$ cells/well). The BCECs $(1x10^5$cells/insert) were seeded in the bottom of the inserts. The membrane of the inserts is made of a microscopically transparent polyester membrane, which is 1.1 cm^2 in diameter and have 0.4 µm (Transwell) or 1 µm sized pores (Millicell), which enables diffusion of molecules through the membrane.

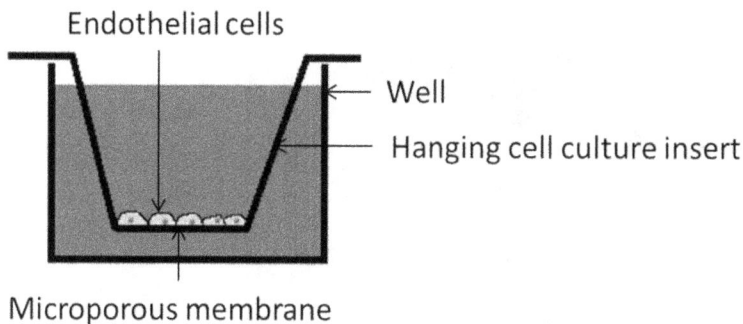

Figure 1 A single well on a 12 well plate with a hanging cell culture insert. The drawing shows a monolayer cultured on the membrane of the hanging insert which is inserted into a well. The BCECs then form a barrier with polarity as *in vivo* with the apical side up and the basolateral side down. This setup can be used as a static *in vitro* BBB model.

DIV-BBB model setup and TEER measurements

The DIV-BBB modules (Flocel Inc.) contain 19 hollow tubes (Figure 2), which have small pores (0.64 µm) to allow diffusion of particles through the walls of the fibers. The experimental setup was adapted from Cucullo et al (2008) [28]. In short the cartridges were coated with collagen on the inside of the hollow fibers and poly-l-lysine on the outside of the fibers. The BCECs were seeded $(1,5x10^5$ cells) on the inside of the fibers and placed in an incubator the first 4 hours without flow to enable attachment. Then astrocytes were seeded $(2x10^5$ cells) on the outside of the hollow fibers. Gas permeable silicone tubes were connected to ports on the upper side of the module for media supply to the cells. The media-flow is operated by a pulsatile pump, which can create a flow rate of 1-50 ml/min. The flow rate was set at 1 ml/min for the first 24-48 hours and bypassed into the outer chamber instead of the lumen of the hollow fibers to allow cell adhesion. Then media was lead into the lumen of the fibers and after 24 hours a sample of the media was taken to count the non-attached BCECs which gives an estimate of how many cells was attached to the hollow fibers. The flow rate was adjusted to 2ml/min the following day and then stepwise with 2 ml/min a day until reaching 6-12 ml/min. In three experiments 550nM/ml hydrocortisone was added to the culture media. The entire setup is placed in a water-jacketed incubator with 5% CO2 at 37°C. Electrodes on the bottom of the module were inserted into a TEER monitoring device, which was directly connected to a computer, providing a curve with the TEER values instantaneously throughout the entire experiment. Before seeding cells in a cartridge a baseline TEER measurement was made on the empty coated cartridge which was subtracted from the TEER values made in the experiment (For further details on the system see flocel.com).

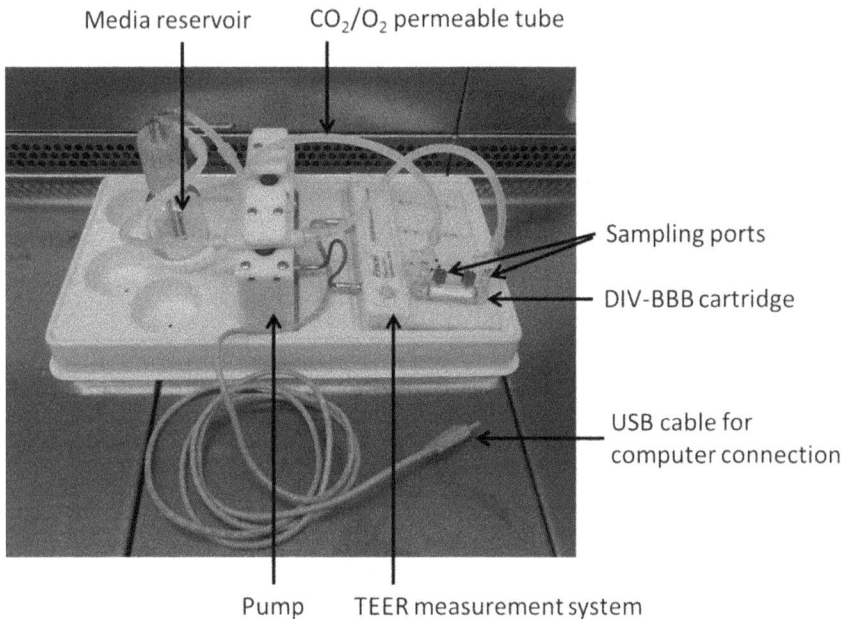

Figure 2 Photo of the DIV-BBB system. The model consists of a cartridge placed in a TEER measurement system. The DIV-BBB cartridge has an inner compartment consisting of 19 hollow fibers made of a microporous membrane. The cartridge has four samplings port and four electrodes for measuring TEER. The cartridge is connected to a media reservoir and a pump by silicone tubing. The TEER measurement system can be connected via USB cable to a computer for instant monitoring of TEER measurements.

TEER measurements of the static BBB models

TEER measurements were conducted with a Millicell™ ERS-2 apparatus (Millipore, USA) and an STX-1 electrode (Millipore). The Millicell ERS-2 is compatible with both the Transwell and Millicell hanging cell culture inserts. The TEER was first measured on an empty insert and subtracted from the TEER values measured on the experimental inserts. The product is then multiplied by the membrane surface area. The Transwell and Millicell inserts both have a membrane area of $1,1 cm^2$.

The TEER was measured every second day the first 5 days and hereafter every day. A steady state would normally commence at day 5-7. Before measurements the culture media was changed and cells and media was allowed to reach room temperature. Three measurements were made on each well from which an average TEER value was calculated.

Immunostaining of the BCECs in the models
This procedure was only conducted on the HBMECs. After the experiment the DIV-BBB cartridges and the hanging culture inserts were washed three times in PBS, fixed in 4% paraformaldehyde for 4 minutes and washed three times in PBS.

The DIV-BBB cartridges were split open and the tubes within were carefully taken out. They were then placed in a 30% sucrose solution. Fibers were stored in glasses in categories of front end fraction, middle fractions and end fractions. The fibers were embedded in Tissue-Tek 4583 O.C.T (Sakura Finetek, Japan) for cryo-sectioning. The fibers were cut in 20 μm thick pieces (Protocol adapted from Cucullo et al (2002) [30]).

The cells in both the inserts and the hollow fiber pieces were incubated overnight with primary antibody, mouse-anti-ZO-1 (Invitrogen) in PBS 1:200 and this was visualized with goat-anti-mouse alexa 488 (Invitrogen) in PBS 1:200. The cell nuclei were stained with DAPI (Sigma-Aldrich) in PBS 1:20 for 5 minutes. The membrane in the hanging cell culture inserts with HBMECs on was cut out of the insert. Both the membranes and the hollow fiber pieces were mounted on a slide with fluorescent moutingmedia (Dako, Denmark) and observed under a fluorescence microscope.

Statistical analysis
An analysis of variance (ANOVA) followed by a Fisher's least significant difference (LSD) method was employed for analyzing the possibility of a difference between the obtained TEER values measured on the four experimental culture setups. A p-value was considered significant at $p<0.05$.

Results and discussion
Many different molecules are manufactured for e.g. treatment of CNS diseases but only a few percent can penetrate the BBB. Testing their penetration abilities *in vivo* on the BBB is both ethically changeling, expensive and time consuming. *In vitro* BBB models have therefore been developed for this purpose. These models mimic the *in vivo* BBB but how good they portray the real BBB functions can be debated. The static *in vitro* BBB model enables the culture of BCECs in co-culture with astrocytes and a brain and blood side can be defined on the BCECs. The static *in vitro* BBB model however lack the ability to support a high TEER and the BCECs are not subjected to a flow which is known to induce higher barrier tightness. Dr. Damir Janigro has together with his group developed a dynamic *in vitro* BBB model which subjects the BCECs to a flow and their studies on the models show

significant increase in TEER values when both primary and immortalized BCECs are grown in this model. In this study the static and dynamic *in vitro* BBB models was tested separately and then compared.

Static in vitro BBB model

The static model was setup in four different ways, 1) monoculture of BCECs, 2) non-contact co culture of astrocytes and BCECs, 3) Contact co-culture of astrocytes and BCECs and 4) contact co-culture of astrocytes and BCECs with hydrocortisone added in the media. TEER measurements made on all four kinds of experimental setups on HBMECs in the hanging cell culture inserts are shown in figure 3. These results resemble the results with RBE4 and Bend3 cells although the TEER values of these cells were approximately 30% lower (data not shown).

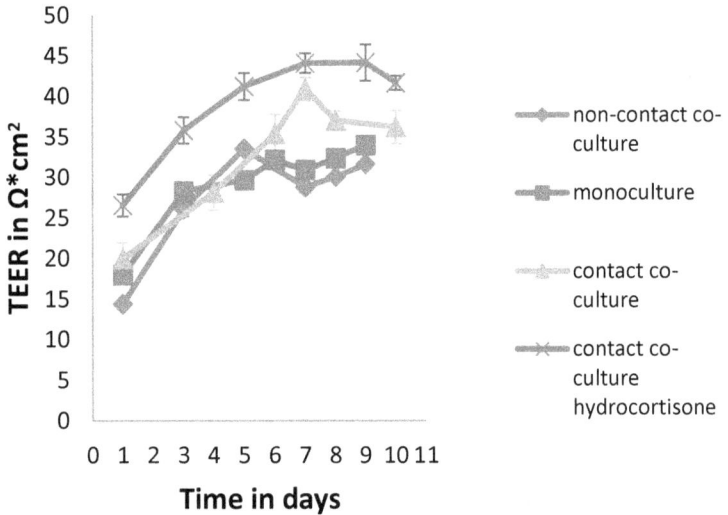

Figure 3 TEER measurements on HBMECs in the static *in vitro* BBB model .The HBMECs grown in the static *in vitro* BBB models did not reach a high TEER compared with in vivo values although a higher TEER is reached when HBMECs are cultured with astrocytes in a contact culture and an even higher TEER if Hydrocortisone is added to the cell media in the contact co-culture (n=22, results presented as means ±standard error).

The TEER values were measured for 9-10 days in total. The TEER values increased until around day 6-8 where it seemed to reach a plateau. Around day 10 the TEER value would slowly begin to decrease again. A statistically significant difference ($p < 0.05$) was found between the TEER threshold value of all of the four setups except between the monoculture and the non-contact co-culture. The test results suggest that astrocytes need to form contact with the BCECs to induce a significant tighter barrier. Furthermore tightness could be further increased if hydrocortisone was added to the media in the contact co-culture setup. This correlates with the findings in Calabria et al (2006) [20]. Their TEER values cannot be directly compared with the values measured in this study as they use primary rat BCECs but they also find that hydrocortisone induce an increase in the TEER threshold values from $94\pm5\Omega*cm^2$ to $218\pm66\Omega*cm^2$ [20]. With the HBMECs the TEER values are very low compared to the ones measured on the primary rat BCECs. HBMECs have been immortalized and passaged for up to 30 passages and therefore have most likely lost some of their barrier functions. It can though be seen on figure 5 that they do stain positive for ZO-1 at the cell borders which indicate that they do form tight junctions, but the low TEER values still imply that there are gaps in between cells.

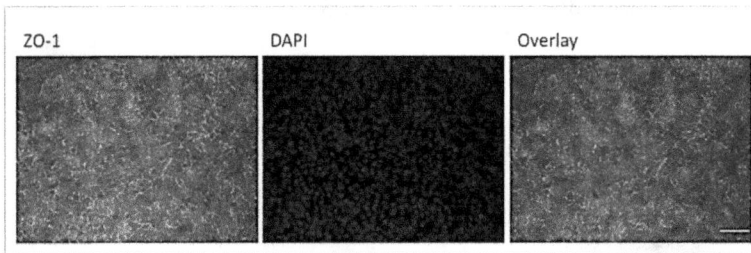

Figure 4 Immunoflourescence staining of HBMECs in a hanging cell culture insert with anti-zo-1 and dapi. The cells stained positive for ZO-1 (scale bar=50μm).

The experimental setup with the hanging cell culture inserts is easy to reproduce and it gives reliable results but the integrity of the barrier does not at all reach the high levels as *in vivo*. One way to optimize this could be to form a BBB of primary BCECs instead of immortalized BCECs. In e.g. Calabria et al 2006 [20] primary rat brain endothelial cells were shown to form tighter barriers in a Transwell system (70-218 $\Omega*cm^2$) [20]. Another approach to mimic the in vivo BBB in a more anatomically correct model is to co culture the BCECs with not just astrocytes but also pericytes. Such a triple cell co-culture model in a Transwell system has been shown to produce higher TEER values. Nakagawa et al (2007,2009) has shown that primary rat brain endothelial cells in a contact co-culture with primary pericytes together with primary astrocytes in remote distance increased the TEER values significantly (~400 $\Omega*cm^2$) compared to mono and double cell cultures (~75-300 $\Omega*cm^2$) [23, 24]. Although the models displays a tighter BBB the TEER values accomplished with the static *in vitro* BBB models with primary rat brain cells in triple co-culture does still not resemble the high TEER values measured *in vivo*. High TEER values have though been measured in static *in vitro* BBB models with primary bovine brain endothelial cells. Helms et al (2010) could produce a barrier with a TEER threshold value of 1638±256 $\Omega*cm^2$ by enhanced media buffer capacity during the growth ofbovine BCECs [31]. A TEER as high as 2100 $\Omega*cm^2$ on primary bovine brain endothelial cells was accomplished by non-contact co-culture with blood-derived macrophages which very much resembles the values obtained *in vivo* [32]. Primary BCECs in mono, double, triple co-culture, addition of hydrocortisone or buffers are all strategies which improve the static in vitro BBB model and make it a useful tool for research on the BBB.

Another approach to accomplish a tighter BBB could be to apply the *in vitro* DIV-BBB model instead which promises a higher TEER and a more correct physiological resemblance.

Dynamic in vitro blood-brain barrier model
Unfortunately it was not succeeded in this study to obtain any reliable results with the DIV-BBB model from Flocel inc. A typical example of TEER measurement of a co-culture in a cartridge is showed in figure 5.

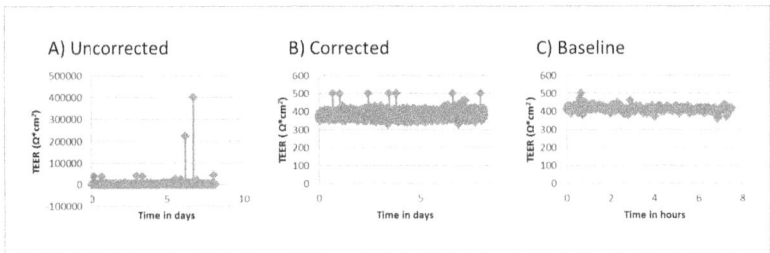

Figure 5 TEER measurements from a cartridge in the DIV-BBB. TEER values from a cartridge with HBMECs cultured in the hollow fibers and HAs cultured on the outside of the fibers are shown in A) uncorrected form, B) corrected form where high and negative values are removed and C) a baseline recording from an empty coated cartridge. The TEER values was recorded automatically every second minute by the TEER measurement device.

The TEER values was automatically recorded every second minute by the TEER measurement device. The TEER values varied a lot and ranged from negative values to five figured values (Figure 5A). Negative and high values are described in the manual of the model to possible be due to air bubbles or clogging of the hollow fibers (flocel.com). Before seeding the cells in the coated cartridges a baseline measurement was obtained (Figure 5C). The baseline value portrays the resistance of the empty cartridge and this value should be subtracted from the experimental TEER values to give the values of the true resistance of the BCECs. In all of the experiments conducted in this study (n=9) with the DIV-BBB the TEER values was not higher than the baseline values as can be seen in the example on figure 5. Here the average baseline TEER was 413 Ω^*cm^2 and the average TEER in the experiment was 381 Ω^*cm^2. Addition of hydrocortisone did not change the TEER values.

The hollow fibers of the DIV-BBB model are made of non-transparent membranes and the cells cultured inside the fibers could therefore not be monitored. To obtain knowledge of the conditions of the hollow fibers and the

BCECs a stain with anti-zo-1 and DAPI was performed. No cells could be identified within the hollow fibers. This could be due to the rather harsh procedure in the immunofluorescence staining protocol which contains several washing steps. It could also be due to cell detachment from the hollow fibers or even no attachment of cell in the fibers to begin with. There was taken some samples from the media to estimate the number of cells that had detached. Approximately 20 % of the loaded cells were found in the media; hence ~80 % were still in the system. There is a possibility that the cells had attached themselves in other parts of the system. Cells was identified on the walls of the cartridges at the entrance and exit points in both ends of the hollow fibers and also in some parts of the tubing. This could explain the lack of an increase in TEER values if the cells did not cover the hollow fibers but instead were attached to other surfaces in the system.

To obtain a tight barrier the BCECs must form a monolayer where they form tight junctions in between them. If there is just a small gap in between two adjacent BCECs in the monolayer the TEER values will be lowered. The surface area of the hollow fibers in the cartridge is quite large ($13,5cm^2$) and therefore it is plausible that it can be difficult to ensure full coverage of the fibers by the BCECs. This could lead to a less loose barrier and very low TEER values as seen in this study.

Comparison of the static and the dynamic in vitro BBB models
When creating an *in vitro* model of the BBB, it is important to obtain values, which mimic the values obtained in living animals. The tightness of the *in vivo* BBB has been poorly reproduced in most studies with the non-dynamic models. The models have been ignoring the fact that the blood flow is a BBB tightness promoting factor. Thus the DIV-BBB model should be a better and more realistic model of the BBB than the static models.

Comparison of the two *in vitro* model types from this study proved difficult as the DIV-BBB model never gave any reliable results. A big difference between the two models is the culture area size and availability. In the static BBB model the membrane on which the BCECs are cultured on is $1,1$ cm^2. The area of the hollow fibers on which the BCECs are cultured on in the static BBB model is $13,5$ cm^2. This is a considerable larger area the BCECs need to cover in the hollow fibers to provide a tight monolayer. Furthermore the culture surface in the dynamic BBB model is in 19 hollow fibers whereas in the static BBB model the surface is flat and horizontal. The membrane in the static model is transparent and the cells are easy to monitor whereas the non-transparent hollow fibers does not allow any cell culture inspections during and experiment. These culture conditions in the static BBB model seems to be more favorable for the BCECs compared to the dynamic BBB model.

At this stage the DIV-BBB model from Flocel Inc. cannot be trusted to produce consistent and reproducible data. It still needs some improvements before it can seriously challenge the more old fashion static *in vitro* BBB models. The static *in vitro* BBB model does not allow shear stress but it still seems to be a reliable model for studies of new carrier compounds. Applying either hydrocortisone to the media or using the triple cell model in the static in vitro BBB model has been shown to reproduce parameters of the *in vivo* BBB in such a way that it can be used as a reliable and trustworthy BBB model.

Acknowledgements

The authors would like to thank Muhammed Hossain and Luca Cucullo at Cleveland Clinic for their constructive input on how to operate the DIV-BBB.
The data in this study were generated by kind grant support from the Danish Medical Research Council (grant no. 271-06-0211), the Spar Nord Fund, and the Obelske Family Fund.

References

[1] J. Lichota, T. Skjørringe, L.B. Thomsen, T. Moos, Macromolecular drug transport into brain using targeted therapy. J. Neurochem, 113 (2010) 1-13.

[2] W. M. Pardridge, Brain drug targeting-The future of brain drug development, Cambridge University Press (2001) Los Angeles, USA

[3] N.J, Abbott, L. Rönnbäck, E. Hansson, Astrocyte-endothelial interactions at the blood-brain barrier, Nat Rev Neurosci, 7 (2006) 41-53

[4] Y. Persidsky, S.H. Ranirez, J. Haorah, G.D. Kanmogne, Blood-brain barrier: structral components and function under physiologic and pathologic conditions. J. Neuroimmune Pharmacol., 1 (2006) 223-36

[5] I. Wilhelm, C. Fazakas, I.A. Krizbai, In vitro models of the blood-brain barrier, Acta Neurobiol exp, 71 (2011) 113-128

[6] E.J. Lee, Y.C. Hung, M.Y. Lee, Early alterations in cerebral hemo-dynamics, brain metabolism, and blood-brain barrier permeability in experimental intracerebral hemorarrhage, J Neurosurg, 91 (1999) 1013-1019

[7] M. Brightman, T.S. Reese, Junctions between intimately apposed cell membranes in the vertebrate brain, J Cell Biol, 40 (1969) 648-677

[8] H.C. Bauer, H. Bauer, A. Lametschwandtner, A. Amberger, P. Ruiz, M. Steiner, Neurovascularization and the appearance of morfological characteristics of the blood-brain barrier in the embryonic mouse central nervous system. Brain Res Dev Brain Res, 75 (1993) 269-278

[9] G.A Grant, N.J. Abbott, D. Janigro, Understanding the physiology of the blood-brain barrier: in vitro models, News Physiol. Sci. 13 (1998) 287-293

[10] L.L. Rubin, D.E. Hall, S. Porter, K. Barbu, C. Cannon, H.C Horner, M. Janatpour, C.W. Liaw, K. Manning, J. Morales, L. Tanner, K.J. Tomaselli, F. Bardet, A cell culture model of the blood brain barrier, J Cell Biol, 115 (1991) 1725-1735.

[11] K.A. Stanness, J.F. Neumaier, T.J. Sexton, G.A. Grant, A. Emmi, D.O. Maris and D. Janigro, A new model of the blood-brain barrier: co-culture of neuronal, endothelial and glial cells under dynamic conditions, Neuroreport ,10 (1999) 3725-3731

[12] J. Correale, A. Villa, Cellular elements of the BBB, Neurochem Res, 34 (2009) 2067-77

[13] M. Krueger, I. Bechmann, CNS pericytes: concepts, misconceptions, and a way out, Glia, 58 (2010) 1-10

[14] R. Daneman, L. Zhou, A.A. Kebede, B.A. Barres, Pericytes are required for BBB integrity during embryogenesis, Nature, 468 (2010) 562-566.

[15] A. Armulik, G. Genove, M. Mäe, M.H. Nisancioglu, E. Wallgard, C. Niaudet, L. He, J. Norlin, P. Lindblom, K. Strittmatter, B.R. Johansson, C. Betsholtz, Pericytes regulate the blood-brain barrier, Nature, 468 (2010) 557-61.

[16] J.M. Tabell, Shear stress and the endothelial transport barrier, 2010, Cardiovasc Res, 2010, 87, 320-330.

[17] M.J. Ott, J.L. Olson and B.J. Ballermann, Chronic in vitro flow promotes ultrastructural differentiation of BCECs, Endothelium 3 (1995) 21-30

[18] S. Santaguida, D. Janigro, M. Hossain, E. Oby, E. Rapp and L. Cucullo, Side by side comparison between dynamic versus static models of blood- brain barrier in vitro: permeability study, Brain Res., 1109 (2006) 1- 13

[19] L. Cucullo, M. Hossain, V. Puvenna, N. Marchi, D. Janigro, The role of shear stress in blood-brain barrier endothelial physiology, BMC Neurosci., 11 (2011) 12:40

[20] A. R. Calabria, C. Weidenfeller, A.R. Jones, H.E Vries, E. V. Shusta, Pyromycin-purified rat brain microvascular endothelial cell cultures exhibit barrier properties in response to glucocorticoid induction, J Neurochem, 97 (2006) 922-933

[21] P.J. Gaillard, L.H. Voorwinden, J.L. Nielsen, A. Ivanov, R. Atsumi, H. Engman, C. Ringbom, A.G de Boer, D.D. Breimer, Establisment and functional

characterization of an in vitro model of the blood-brain barrier, comprising a co-culture of brain capillary BCECs and astrocytes, Eur J Pharm Sci, 12 (2001) 215-222.

[22] M.A. Deli, C.S. Ábrahám, Y. Kataoka, M. Niwa, Permeability studies on in vitro blood-brain barrier models: Physiology, pathology, and pharmacology, Cell Mol Neurobiol, 25 (2005) 59-127.

[23] S. Nakagawa, M.A. Deli, S. Nakao, M. Honda, K. Hayashi, R. Nakaoke, Y, Kataoka, M. Niwa, Pericytes from brain microvessels strengthen the barrier integrity in primary cultures of rat brain endothelial cells, Cell Mol Neurobiol, 27 (2007) 687-694

[24] S. Nakagawa, M.A. Deli, H. Kawaguchi, T. Shimizudan, T. Shimono, A. Kittel, K. Tanaka, M. Niwa, A new blood-brain barrier model using primary rat brain endothelial cell, pericytes and astrocytes, Neurochem Int, 54 (2009) 253-263

[25] H. Vernon, K. Clarck, J.P Bressler, In vitro models to study the blood brain barrier, In vitro neurotoxicology: Methods and protocols, Chapter 10, 758 (2010) 153-168.

[26] L. Cucullo, B. Aumayr, E. Rapp, D. Janigro, Drug delivery and in vitro models of the blood-brain barrier, Curr. Opin. Drug Discov. Devel. 8 (2005) 89-99

[27] L. Cucullo, M. Hossain, E. Rapp, T. Manders, N. Marchi, D. Janigro, Development of a humanized in vitro blood-brain barrier model to screen for brain penetration of antiepileptic drugs, Epilepsia, 48 (2007) 505-516

[28] L. Cucullo, P.O. Couraud, B. Weksler, I.A. Romero, M. Hossain, E. Rapp, D. Janigro, Immortalized human brain BCECs and flow-based vascular modeling: a marriage of convenience for rational neurovascular studies, J Cereb Blood Flow Metab., 28 (2008) 312-328

[29] M. Hossain, T. Sathe, V. Fazio, P. Mazzone, B. Weksler, D. Janigro, E. Rapp, L. Cucullo, Tobacco smoke: A critical etiological factor for vascular impairment at the blood-brain barrier, Brain Res., 1287 (2009) 192-205

[30] L. Cucullo, M.S. McAllister, K. Knight, L. Krizanac-Bengez, M. Maroni, M.R. Mayberg, K. A. Staness, D. Janigro, A new dynamic in vitro model for the multidimensional study of astrocyte-endothelial cell interactions at the blood-brain barrier, Brain Res. 951 (2002) 243-254

[31] H.C. Helms, H.S. Waagepetersen, C.U. Nielsen, B. Brodin, Paracellular tightness and claudin-5 expression is increased in the BCEC/astrocyte blood-brain

barrier model by increasing media buffer capacity during growth, AAPS J, 12 (2010) 759-770

[32] D. Zenker, D. Begley, H. Bratzke, H. Rübsamen-Waigmann, H. von Briesen, Human blood-derived macrophages enhance barrier function of cultured brain capillary BCECs, J Physiol, 551 (2003) 1023-1032

4 Discussion

The overall objective of this study was to find potent drug carriers for either direct delivery to the brain or delivery of cDNA to BCECs with the potential of de novo gene expression and subsequent secretion of synthesized proteins to the brain. The aim was also to establish a valid *in vitro* BBB model for testing these drug carriers.

4.1   PULLULAN-SPERMINE COMPLEXES AS GENE-CARRIERS TO BCECS

Pullulan-spermine was successfully conjugated with plasmid cDNA encoding the red fluorescent protein, HcRed-1. The formed polyplex consisting of pullulan-spermine-pHcRed-1 cDNA was introduced to the BCECs and transgene BCECs expressing the red fluorescent marker was detected. Pullulan-spermine complexes were also conjugated with plasmid cDNA encoding hGH1 and transfection of the BCECs was further confirmed by the expression of hGH1 mRNA by BCECs. The results clearly show that pullulan-spermine is a potent carrier of genetic material suitable for transfection of BCECs, as it succeeded in delivering its cargo to the cell cytosol with a subsequent transport to the cell nucleus. These findings supports results from other studies in where pullulan-spermine proved to be a potent donor of genetic material to cells of non-neuronal origin like human bladder cancer cells (T24), human hepatoma cells (HepG2), and mesenchymal stem cells [64, 65, 67, 68, 69]. The results of the present study add to this row of data via the discovery that BCECs can be converted into protein factories for protein secretion to the brain following uptake and transfection of cDNA carried into the cells by pullulan-spermine. In this thesis, gene therapy was performed with plasmid cDNA encoding hGH1 but theoretically, this principal method for cellular tranfection might be used for delivery of any protein of relevance for the brain. Jiang and co-workers (2003) transfected cultured mouse brain capillary endothelial cells (MBEC4) with pIRESneo-mGDNF by using Lipofectamine following secretion of GDNF [110]. They also proved secretion of GDNF by the MBEC4 cells to both the apical and the basolateral side of the MBEC4 cells. Furthermore they were able to transfect BCECs in vivo with GDNF encapsulated in Hemagglutination virus of Japan (HVJ)-liposomes and the secreted GDNF provided neuroprotection for dopamine neurons against 6-hydroxydopamine induced lesions [110]. This supports the possibility of using BCECs as protein secreting factories for secretion of proteins that could have a beneficial effect on damaged neurons or other cell types in the brain. BCECs could potentially be transfected to secrete other proteins than hGH1. GDNF and BDNF have been shown to play a significant role in maintenance of fully differentiated neurons and to promote growth and differentiation of newly formed neurons [111, 112]. Likewise EPO, bFGF, and NGF are also of putative interest (see Table 2 in Introduction). Johnston et al (1996) showed that bFGF could be transferred into BCECs *in vitro* by liposome complexes and subsequent secretion of bFGF into the culture media by the BCECs was detected [113]. It should also be mentioned that secretion of hGH1 may have a beneficial effect on not only neurons but also oligodendrocytes and astrocytes [114].

In this study immortalized BCECs grown in monolayer was employed for transfection with pullulan-spermine-cDNA complexes. However, there can be a big difference between the barrier properties and other characteristics of immortalized and primary BCECs and experiments should preferably be performed on polarized primary BCECs to increase the significance on how the results could translate to the *in vivo* situation with respect to synthesis and secretion. Jiang and co-workers showed a huge potential for basolateral secretion of protein from BCECs which should also be investigated by the use of the vector of the present study [110].

4.2   TRANSPORT OF PULLULAN-SPERMINE CARGO INTO THE CELL NUCLEUS

Presumably, gene delivery by pullulan-spermine is limited to mitotic cells [65]. One of the new findings in this thesis was that BCECs present in either a dividing or non-dividing state could be transfected by pullulan-spermine-cDNA complexes, which suggests that the plasmid cDNA not only enters the cell nucleus during mitosis but are also trafficked through the intact nuclear membrane during the non-mitotic state of the cell cycle. This trafficking is though not the main route for plasmid cDNA as the transfection in non-dividing cells is lower than in dividing cells.

The transport of plasmid cDNA to the cell nucleus was demonstrated using coupling of NLSs to plasmid cDNA, which led to increase in *in vivo* transfection (e.g. [50, 51, 115]). Possibly the coupling of NLSs to the pullulan-spermine-plasmid cDNA complex would increase delivery of cDNA to the nucleus of non-mitotic cells. Plasmid cDNAs used in this study encode HcRed-1 and hGH1 and contain sequences of SV40 and cytomegalovirus (CMV) respectively. SV40 contains NLSs and is able to optimize nuclear uptake whereas CMV does not facilitate such nuclear translocation [116]. A change in the plasmid vector composition of the hGH1 cDNA might also increase its nuclear uptake and thereby further strengthen the carrier properties of pullulan-spermine-plasmid cDNA complexes.

4.3   TARGETING PROPERTIES OF PULLULAN-SPERMINE

The intracellular route by which pullulan-spermine complexes are internalized by BCECs was not investigated in the present study, however other studies have shown that positively charged polyplexes can undergo non-specific adsorptive endocytosis via interaction with anionic proteoglycans and glycoproteins present on the luminal cell surface [60, 61, 62]. Specifically pullulan-spermine is thought to be taken up by cells via sugar-recognition receptors [65]. The pullulan-spermine-DNA complexes employed in this study were 240-300nm in size and therefore attributable to enter the BCECs by means of calveolae dependent endocytosis [65]. As either of these mechanisms for recognition and uptake is thought to be specific for BCECs *in vivo* a targeting strategy for BCEC access is necessary.

One strategy could be to administer the pullulan-spermine complex directly into the carotid artery which would increase the rate of interaction between the BCECs and the complex before interaction with other cells. This will of course not exclude interaction with other cell types but as the complexes would pass the BCECs immediately after injection the charged complexes would be prone to interaction with the BCECs before encountering other organs. A downside to this strategy would be that the pullulan-spermine complexes would still be distributed into the systemic circulation outside CNS and this could give rise to unwanted side-effect in non-target organs.

Another approach could be to target a receptor on the apical surface of the BCECs. It has been shown that OX26 will bind to the transferrin receptor which leads to internalization of the antibody into the cytosol of BCECs when administered to rats intravenously or by in situ perfusion [41, 42]. This makes OX26 a suitable targeting molecule to the BCECs.

The conjugation of OX26 to the pullulan-spermine-cDNA complex was done successfully suggesting that this principle form of targeting strategy is accessible (Lichota et al (unpublished data)). This strategy will though still not exclude uptake by other cell types expressing the transferrin receptor, but by administration into the carotid artery the possibility of uptake mainly by BCECs would increase.

4.4   PULLULAN-SPERMINE AND SERUM COMPATIBILITY

The transformation efficiency of pullulan-spermine-cDNA complexes is severely inhibited by serum [68] a finding supported by data of the present study. Hence, almost no transgenic BCECs expressing HcRed1 C1 were detected when serum was added to the growth media. This lack of transfection is thought to be due to the negatively charged serum proteins which lower or even completely neutralize the charge of the cationic polyplexes. This apparent obstacle for future in vivo experiments was addressed by Thakor and co-workers who developed a strategy where pullulan-spermine-cDNA complexes would form anionic complexes, anioplexes [57, 117]. These anioplexes have proved significantly more effective for transfection than their cationic counterparts; the rationale being that anionic serum proteins will not interact with the anioplexes, and therefore serum is no longer a restraining factor for the complex as the interaction with the negatively charged cell surface components is no longer taking place [57, 117]. An apparent disadvantage in this strategy is that it limits non-specific endocytic uptake, which must be dealt with by making the complexes targetable as described in the previous section. This targeting approach may in fact prove to be advantageous as the ratio of specific to unspecific uptake probably will be markedly improved.

Another strategy to avoid serum protein interaction could be to encapsulate the pullulan-spermine-cDNA complexes in liposomes. Liposomes have been very intensely studied for their capabilities as drug-carriers. They can be PEGylated and conjugated with antibodies [118] suggesting that PEGylated targetable liposomes carrying pullulan-spermine-cDNA could denote a potent complex being both serum compatible and targetable.

4.5   SPIOS AND DRUG DELIVERY TO THE BRAIN

The uptake and transport of SPIOs through BCECs indicates that they are appropriate candidates for drug delivery to both BCECs and the brain. Their passage through BCECs occurred in small scale without external aid. Application of an external magnetic field clearly enhanced the SPIOs movement through the BCECs. Once through BCECs cultured in cell culture inserts, the SPIOs were taken up by astrocytes, grown in wells in which the inserts with BCECs were placed.

The SPIOs used in the present study is coated with starch which have terminal hydroxyl groups. These functional hydroxyl groups can be covalently coupled with amine groups on antibodies or other types of proteins. As SPIOs can also be coated with substrates like chitosan or phospholipids with capabilities to bind ligands, cDNA and drugs [80, 82, 84, 85], the SPIOs are potent drug carriers to the brain in a targetable manner. SPIOs coated with chitosan and enclosed in liposomes are able to carry plasmid DNA into BCECs and transfect them in culture (Linemann et al (unpublished data)).

4.6   SPIOS AND POSSIBLE DAMAGING EFFECTS

Accumulating an excess of iron present as iron oxide particles could potentially become a safety issue due to the risk of metal-induced cytotoxicity and damage to the BCECs [80, 85]. Not only could BCECs be damaged but they could also lose their integrity to proteins in circulation leading to increased BBB permeability.

No significant damage was seen in BCECs after having been subjected to SPIOs and an external magnetic field, as less than one 1% of the cells were damaged. Iron oxide SPIOs coated with tween 80 and of a diameter of 30 nm have been shown to be cytotoxic to murine macrophages (J774) after incubation for 6 hours at a concentration of 200µg/ml, but not significantly toxic at 100µg/ml [79]. The present study shows that iron oxide SPIOs of a diameter of ~117,5nm is non-toxic to the BCECs at a concentration of 140µg/ml and 5 hour incubation time. Additionally, the TEER measurements implied no disruption of the BBB following application of SPIOs, the external magnetic field or both. Hence, the integrity of the BBB remains stable after application and passage of the SPIOs indicating that magnetic particles would be suitable also for *in vivo* studies.

4.7   SPIOS AND MAGNETIC FORCE-TARGETED DELIVERY

The external magnetic field was supplied by a plate magnet with a strength of 0.39 Tesla, which is compatible with the strength of the magnetic fields applied for magnetic resonance imaging (MRI) in clinic. MRI is normally powered by 0.2 - 3 Tesla with the most common values being in the range of 1,5 and 3 Tesla, but in some analyzes MRI may be performed at 30 Tesla. As SPIOs are already in clinical use as contrast agents for MRI scanning [71] is plausible that delivery to the brain of drugs and genes carried by SPIOs is accessible even with simultaneously real-time visualization of their accumulation in the brain.

4.8   STATIC VERSUS DYNAMIC *IN VITRO* BBB MODEL

Unfortunately it was not possible to obtain any reliable results with the dynamic *in vitro* BBB model in the present study, which makes it impossible to compare the static and dynamic BBB models. Other research groups have proven that it is possible to obtain data from the dynamic model and their studies were the reason for investing in this model [100, 101, 102, 103, 104, 105, 106, 107, 108, 109].

In the dynamic model there are a lot of small finesses that can alter the culturing process. The rather large surface area is limiting because BCECs needs to cover the entire surface before an increased TEER can be measured. In the static *in vitro* BBB model the surface area to be covered by BCECs is significantly smaller. Therefore for practical reasons the culture conditions of the static model favors the establishment of a tight BBB with high TEER values.

In the dynamic BBB model the cell culturing is very difficult to monitor and therefore very difficult to get a comprehensive view on whether BCECs form a confluent monolayer. The hollow fibers are not transparent like the microporous membranes of the static model. It is therefore also very difficult to ensure correct loading of the BCECs into the hollow fibers or to monitor the growth. These problems could easily be changed, by forming hollow fibers of transparent material, instead of the current non-transparent membranes.

It was also difficult to avoid unwanted attachment of cells on surfaces outside hollow fibers. If the two sampling ports to the inner compartment were exclusively attached to the hollow fibers this unwanted attachment of cells in the cartridge could be avoided. Unwanted attachment of cells is not an issue in the static model, and although the static model may seem simpler and lack some key features for induction of the BBB phenotype, it remains the best model for obtaining reproducible data on BCECs.

4.9   IMMORTALIZED BCECS AND BBB INTEGRITY

When studying transcellular or intercellular transport of substances into BCECs it is important to eliminate paracellular leakage. The BCECs of the present study did not express too impressive TEER values. Although they stained positive for ZO-1 the low TEER values indicate that the BCECs are not that closely interconnected by tight junctions. ZO-1 formation indicates tight junction formation but it is a cytoplasmic plaque protein that links the transmembrane junctional proteins to the actin cytoskeleton. Staining for the transmembrane tight junction proteins e.g. occludin or claudin-5 could have provided a more clear view of the presence of tight junctions. Furthermore a study on the barrier permeability with a tracer e.g. sodium fluorescein or radiolabeled sucrose could have been performed. The permeability depends on the sum of transport across all junctional pathways [87] and would therefore provide insight into the tightness of the barrier as well. The BCECs in the present study have probably lost some of their barrier characteristics during their repetitive passaging, but their BBB properties could be increased by a contact co-culture with astrocytes and further increased by addition of hydrocortisone, observations in hand with other laboratories (e.g. [88, 94, 96, 98]). Despite attempt to optimize the culture conditions the HBMEC cell line used in the present study does not seem to be able to form a tight enough barrier to fulfill the criteria of TEER values around 150-200 which have been determined to be necessary for obtaining reasonable information from an *in vitro* BBB model [87, 119].

Primary brain microvascular BCECs could instead be employed to improve the *in vitro* model of the BBB. Primary BCECs have intact BBB features and form a much tighter barrier than their immortalized counterparts [86, 87]. TEER values measured on primary cultures of e.g. bovine and porcine BCECs can be as high as values obtained in vivo [87, 120, 121]. Including both astrocytes and pericytes in double or triple co-culture with the BCECs has also been shown to strengthen the tightness of the BBB models [95, 122]. With the use of primary BCECs in monoculture or coculture with astrocytes and/or pericytes, the *in vitro* BBB models are more compatible with the *in vivo* situation and therefore more useful for studying the passage of various compounds as e.g. SPIOs and pullulan-spermine-cDNA through the *in vitro* BBB.

5. Future Perspectives

In this study it was found that pullulan-spermine complexes are potent gene carriers to BCECs. SPIOs were found to be potentially potent carriers for delivery to both BCECs and the brain. Furthermore it was found that the static in vitro BBB model consisting of cell culture inserts were the most reliable BBB model when compared with a dynamic in vitro BBB model. The results in this thesis have raised three new questions I would like to be able to answer in the nearer future.

1) First, could a better *in vitro* BBB model be established based on the static model? The three studies in this thesis might benefit from establishment of a primary brain endothelial cell culture. Primary BCECs possess far more of the BBB characteristics, and especially they form a tighter barrier than the immortalized BCECs [87]. The existing literature suggests that replacement of immortalized BCECs with primary cells would highly improve the results obtained in the static in vitro BBB model and make them translate to the *in vivo* situation [87]. Attempts have already been made on establishing human primary BCECs during the last period of this Ph.D. study. Pieces of human brain tissue are obtained from patients undergoing surgery to remove brain tumors at the neurosurgical department on the Hospital of Aalborg. Unfortunately it has been a challenge to ensure a pure fraction of BCECs, and the protocol still needs further improvement. Human brain tissue is not provided on a regular basis, therefore more available sources as for example rat brains should be studied as well.

2) Secondly, is it possible to alter pullulan-spermine complexes to become potent gene-carriers for *in vivo* use? If pullulan-spermine complexes are to be employed for *in vivo* purposes the serum incompatibility problem has to be solved. Pilot studies have been initiated to combine PEGylated liposomes with pullulan-spermine employing a new protocol for liposome preparation [123]. The hypothesis to examine is that if PEGylated liposomes can carry pullulan-spermine-cDNA complexes into BCECs, release pullulan-spermine-cDNA into the cytosol from where cDNA will reach to the nucleus to enable transfection. This strategy ensures full protection of the pullulan-spermine complexes from serum degradation. Another strategy would be to form anionic complexes of pullulan-spermine-DNA as recently described by Thakor et al (2011) [117], which was shown to eliminate the serum incompatibility factor.

3) Thirdly, is it possible to conjugate the fluorescent SPIOs with a cargo and demonstrate delivery of this cargo into BCECs or directly into the brain?

Having studied the ability of SPIOs to enter into BCECs and even through the BCECs, the next step would be to conjugate these coated particles with both cDNA and a BCEC targetable molecule. DNA binding could be done by coating with a positively charged molecules e.g. chitosan to take advantage of the electrostatic binding ability. Furthermore the magnetic particles can be covalently coupled to antibodies via cyanogen bromide activation, which should be tested as a strategy towards producing BCEC specific targetable SPIOs.

References

[1] Pardridge, W.M., Introduction to the blood-brain barrier: Methodology, Biology and Pathology, *Cambridge University Press*, 2006.

[2] Lichota, J., Skjørringe, T., Thomsen, L.B., Moos, T., Macromolecular drug transport into brain using targeted therapy. *J. Neurochem*, Vol. 113, 2010, pp. 1-13.

[3] Abbott, N.J., Rönnbäck, L., Hansson, E., Astrocyte-endothelial interactions at the blood-brain barrier, *Nat Rev Neurosci*, Vol. 7, 2006, pp. 41-53.

[4] Correale, J. and Villa, A., Cellular elements of the BBB, *Neurochem Res*, Vol. 34, 2009, pp. 2067-77.

[5] Erickson, A.C. and Couchman, J.R., Still more complexity in mammalian basement membranes, *J. Histochem Cytochem*, Vol. 48, 2000, pp. 1291-1306.

[6] Abbott, N.J., Patabendige, A.A.K., Dolman, D.E.M., Yusof, S.R., Begley, D.J., Structure and function of the blood-brain barrier. *Neurobiol. Dis*. Vol. 37, 2010, pp. 13-25.

[7] Persidsky, Y., Ranirez, S.H., Haorah, J., Kanmogne, G.D., Blood-brain barrier: structral components and function under physiologic and pathologic conditions. *J. Neuroimmune Pharmacol*., Vol. 1, 2006, pp. 223-36.

[8] Krueger, M. and Bechmann, I., CNS pericytes: concepts, misconceptions, and a way out, *Glia*, Vol. 58, 2010, pp. 1-10.

[9] Daneman, R., Zhou, L., Kebede, A.A., Barres, B.A., Pericytes are required for BBB integrity during embryogenesis, *Nature*, Vol. 468, 2010, pp. 562-566.

[10] Armulik, A., Genove, G., Mäe. M., Nisancioglu, M.H., Wallgard, E., Niaudet, C., He, L., Norlin, J., Lindblom, P., Strittmatter, K., Johansson, B.R., Betsholtz, C., Pericytes regulate the blood-brain barrier, *Nature*, Vol 468, 2010, pp. 557-61.

[11] Lee, E.J., Hung, Y.C, Lee, M.Y., Early alterations in cerebral hemo-dynamics, brain metabolism, and blood-brain barrier permeability in experimental intracerebral hemorarrhage, *J Neurosurg*, Vol. 91, 1999, pp. 1013-1019.

[12] Pardridge, W.M., The blood-brain barrier: bottleneck in brain drug development, *NeuroRX*, Vol. 2, 2005, pp. 3-14.

[13] Brightman, M. and Reese, T.S., Junctions between intimately apposed cell membranes in the vertebrate brain, *J Cell Biol*, Vol. 40, 1969, pp. 648-677.

[14] Bauer, H.C., Bauer, H., Lametschwandtner, A., Amberger, A., Ruiz, P., Steiner, M., Neurovascularization and the appearance of morfological characteristics of the blood-brain barrier in the embryonic mouse central nervous system, *Brain Res Dev Brain Res.*, Vol 75, 1993, pp. 269-278.

[15] Begley, D.J., Structure and function of the blood-brain barrier, In: Touitou, E; Barry, B.W. Enhancement in drug delivery, *CRS Press, Boca Raton*, 2007, pp. 575-591.

[16] Stewart, P.A., Endothelial vesicles in the blood-brain barrier: are they related to permeability? *Cell Moll Neurobiol*, Vol. 20, 2000, pp. 149-163.

[17] Begley, D.J. and Brightman, M.W., Structural and functional aspects of the blood-brain barrier, *Prog Drug Res*, Vol. 61, 2003, pp. 39-78.

[18] Begley, D.J., ABC transporters and the blood-brain barrier, *Curr Pharm Des*, Vol. 10, 2004, pp. 1295-1312.

[19] Pardridge, W.M., Brain drug targeting, the future of brain drug development, *Cambridge University Press*, 2001.

[20] O'Reilly, M.A., Huang, Y., Hynynen, K., The impact of standing wave effects on transcranial focused ultrasound disruption of the blood-brain barrier in a rat model, *Phys Med Biol*, Vol. 55, 2010, pp. 5251-5267.

[21] Etu, J., Wang, M., Joshi, S., Enhanced disruption of the blood-brain barrier by intrcarotid mannitol injection during transient cerebral hypoperfusion in rabbits, *J Neurosurg Anestesiol*, Vol. 19, 2007, pp. 249-256.

[22] Black, K.L. and Ningaraj, N.S., Modulation of brain tumor capillaries for enhanced drug delivery selectively to brain tumor, *Cancer Control*, Vol. 11, 2004, pp. 165-173.

[23] Lu, C., Diehla, S.A., Noubadea, R., Ledouxb, J., Nelsonb, M.T., Spacha, K., Zacharyc, J.F., Blankenhornd, E.P., Teuscher, C., Endothelial histamine H1 receptor signaling reduces blood–brain barrier permeability and susceptibility to autoimmune encephalomyelitis, *PNAS*, Vol. 107, 2010, pp. 18967-18972.

[24] Huynh, G.H.; Deen and D.F.; Szoka, Jr. F.C., Barriers to carrier mediated drug and gene delivery to brain tumors, *J Control Release*, Vol. 110, 2006, pp. 236-259.

[25] Mahoney, M.J. and Saltman, W.M., Controlled release of proteins to tissue transplants for treatments of neurodegenerative disorders, *J Pharm Sci*, Vol. 85, 1996, pp. 1276-1281.

[26]   Alam, M.I., Beg, S., Samad, A., Baboota, S., Kohli, K., Ali, J., Ahuja, A., Akbar, M., Strategy for effective brain drug delivery, *Eur J Pharm Sci*, Vol. 40, 2010, pp. 385-403.

[27]   Pardridge, W.M., Blood-brain delivery, *Drug Discov Today*, Vol. 12, 2007, pp. 54-61.

[28]   Pardridge, W.M., Overcoming the blood-brain barrier, *Mol Interv*, 3, 2003, pp. 90-105.

[29]   Dubey, R.K., McAllister, C.B., Inoue, M., Wilkinson, G.R., Plasma binding and transport of diazepam across the blood-brain barrier: No evidence for in vivo enhanced dissociation, *J Clin Invest*, Vol. 84, 1989, pp. 1155-1159.

[30]   Friedmann, T. and Roblin, R., Gene therapy for human genetic disease, *Science*, Vol. 175, 1972, pp. 949-955.

[31]   Kawakami, S., Higuchi, Y., Hashida, M., Non-viral approaches for non-targeted delivery of plasmid DNA and oligonucleotides, *J Pharm Sci*, Vol. 97, 2008, pp. 726-745.

[32]   Pathak, A., Patnaik, S., Gupta, K.C., Recent trends in non-viral vector-mediated gene delivery, *J Biotechnol*, Vol. 4, 2009, pp. 1559-1572.

[33]   Nishikawa, M. and Huang, L., Nonviral vectors in the new millenium: Delivery barriers in gene transfer, *Hum Gene Ther*, Vol. 12, 2001, pp. 861-870.

[34]   Biju, K.C., Zhou, Q., Li, G., Imam, S.Z., Roberts, J.L., Morgan, W.W., Clark, R.A., Li, S., Macrophage-mediated GDNF delivery protects against dopaminergic neurodegeneration: A therapeutic strategy for parkinson's disease, *Mol Ther*, Vol. 8, 2010, pp 1536-1544.

[35]   Kawabata, K., Takakura, Y., Hashida, M., The fate of plasmid DNA after intravenous injections in mice: involvement of svavenger receptors in its hepatic uptake, *Pharm Res*, Vol. 12, 1995, pp. 825-830.

[36]   Lentz, T.B., Gray, S.J., Samulski, R.J., Viral vectors for gene delivery to the central nervous system, *Neurobiol Dis*, 2011, in press.

[37]   Thaci, B., Ulasov, I.V., Wainwright, D.A., Lesniak, M.S., The challenge for gene therapy: Innate immune responses to adenovirus, *Oncotarget*, Vol. 2, 2011, pp. 113-121.

[38]   He, C., Tabata, Y., Gao, J., Non-viral gene delivery carrier and its three dimensional transfection system, *Int J Pharm*, Vol. 386, 2010, pp. 232-242.

[39]   Jeffries, W.A., Brandon, M.R, Hunt, S.V., Williams, A.F., Gatter, K.C., Mason, D.Y., Transferrin receptor on endothelium of brain capillaries, *Nature*, Vol. 312, 1984, pp. 162-163.

[40]   Frieden, P.M., Walus, L.R., Musso, G.F., Taylor, M.A., Malfroy, B., Starzyk, R.M., Anti-transferrin receptor antibody and antibody drug

conjugates cross the blood-brain barrier, *Proc Natl Acad Sci USA*, Vol. 88, 1991, pp. 4771-4775.

[41]   Moos, T. and Morgan, E.H., Restricted transport of anti-transferrin receptor antibody (OX26) through the blood-brain barrier in the rat, *J Neurochem*, Vol. 79, 2001, pp. 119-129.

[42]   Gosk, S., Vermehren, C., Storm, G., Moos, T., Targeting anti-transferrin receptor antibody (OX26) and OX26-conjugated liposomes to brain capillary endothelial cells using in situ perfusion, *J Cereb Blood Flow Metab*, Vol. 24, 2004, pp. 1193-1204.

[43]   Ponka, P. and Lok, C.N., The transferrin receptor: role in health and disease, *Int J Biochem Cell Biol*, Vol. 31, 1999, pp. 1111-1137.

[44]   Anderson, G.J. and Vulpe, C.D., Mammalian iron transport, *Cell Mol Life Sci*, Vol. 66, 2009, pp. 3241-3261.

[45]   Nel, A.E., Mädler, L., Velegol, D., Xia, T., Hoek, E.M.V., Somasundaran, P., Klaessig, F., Castranova, V., Thompson, M., Understanding biophysicochemical interactions at the nano-bio interface, *Nat Mater*, Vol. 8, 2009, pp. 543-557.

[46]   Lechardeur, D., Sohn, K.J., Haardt, M., Joshi, P.B., Monck, M., Graham, R.W., Beatty, B., Squire, J., O'Brodovich, H., Lukacs, G.L., Metabolic instability of plasmid DNA in the cytosol: a potential barrier to gene transfer, *Gene Ther*, Vol. 6, 1999, pp. 482-497.

[47]   Feldherr, C.M. and Akin, D., The location of the transport gate in the nuclear pore complex. *J Cell Sci*, Vol. 10, 1997, pp. 3065-3070.

[48]   Dean, D.A., Import of plasmid DNA into the nucleus is sequence specific, *Exp Cell Res*, Vol. 230, 1997, pp. 293-302.

[49]   Wangstaff, K.M. and Jans, D.A., Nucleocytoplasmatic transport of DNA: enhancing non-viral gene transfer, *Biochem J*, Vol. 406, 2007, pp. 185-202.

[50]   Zanta, M.A., Belguise-Valladier, P., Behr, J.P., Gene delivery: a single nuclear localization signal peptide is sufficient to carry DNA to the cell nucleus, *Prod Natl Ascad Sci USA*, Vol. 96, 1999, pp. 91-96.

[51]   Subramanian, A., Ranganathan, P., Diamond, S.L., Nuclear targeting peptide scaffolds for lipofection of nondividing mammalian cells, *Nat Biotechnol*, Vol. 17, 1999, pp. 873-7.

[52]   Zhang, H. and Vinogradov, S.V., Short biodegradeble polyamines for gene delivery and transfection of brain capillary BCECs, *J. Control. Release*, Vol. 143, 2010, pp. 359-66.

[53]   Brown, M.D., Schatzelein, A., Brownlie, A., Jack, V., Wang, W., Tetley, L., Gray, A.I., Uchegbu, I.F., Preliminary characterization of novel amino acid based polymeric vesicles as gene delivery agents, *Bioconjug Chem*, Vol. 11, 2000, pp. 880-891.

[54] Ishii, T., Okahata, Y., Sato, T., Mechanism of cell transfection with plasmid/chitosan complexes, Biochim. Biophys. Acta, Vol. 1514, 2001, pp. 51-64.

[55]

[56] Barnjee, P., Reichardt, W., Wiessleder, R., Bogdanov, A.Jr., Novel hyperbranced Dendron for gene transfer in vitro and in vivo, Bioconj Chem, Vol. 15, 2004, pp. 960-968.

[57] Thakor, D.K., Teng, Y.D., Tabata, Y., Neuronal gene delivery by negatively charged pullulan-spermine/DNA anioplexes, Biomaterials, Vol. 30, 2009, pp. 1815-1826.

[58] Dash, P.R., Read, M.I., Barrett, L.B., Wolfert, M.A., Seymour, L.W., Factors affecting blood clearance and in vivo distribution of polyelectrolyte complexes for gene delivery, Gene Ther, Vol. 6, 1999, pp. 643-650.

[59] Caliceti, P. and Veronese, F.M., Pharmacokinetic and biodistribution properties of poly(ethyleneglycol)-protein conjugates, Adu Drug Deliv Rev, Vol. 55, 2003, pp. 1261-1277.

[60] Ruponen, M., Rönkkö, S., Honkakoski, P., Pelkonen, J., Tammi, M., Urtti, A., Extracellular glycosaminoglycans modify cellular trafficking of lipoplexes and polyplexes, J Biol Chem, Vol. 276, 2001, pp. 33875-33880.

[61] Boussif, O., Lezoualch, F., Zanta, M.A., Mergny, M.D., Scherman, D., Demeneix, B., Behr, J.P., A versatile vector for gene and oligonucleotide transfer into cells in culture and in vivo: polyethylenimine, Proc Nat Acad Sci USA, Vol. 92, 1995, pp. 7297-7301.

[62] Moghimi, S. M., Symonds, P., Murray, J.C., Hunter, A.C., Debska, G., Szewczyk, A., A two-stage poly(ethylenimine)-mediated cytotoxicity: implications for gene transfer/therapy. Mol. Ther., Vol. 11, 2005, pp. 990-95.

[63] Pichon, C., Conçalves, C., Midoux, P., Histidine rich peptides and polymers for nucleic acid delivery, Adv Drug Deliv Rev, Vol. 53, 2001, pp. 75-94.

[64] Jo, J., Ikai, T., Okazaki, A., Yamamoto, M., Hirano, Y., Tabata, Y., Expression profile of plasmid DNA by spermine derivatives of pullulan with different extents of spermine introduced, J. Control. Release, Vol. 118, 2007a, pp. 389-98.

[65] Kantani, I., Ikai, T., Okazaki, A., Jo, J., Yamamoto, M., Imamura, M., Kanematsu, A., Yamamoto, S., Ito, N., Ogawa, O., Tabata, Y., Efficient gene transfer by pullulan-spermine occurs through both clathrin- and raft/calveolae-dependent mechanisms, J Control Release, Vol. 116, 2006, pp. 75-82.

[66]   Hosseinkhani, H., Aoyama, T., Ogawa, O., Tabata, Y., Liver targeting of plasmid DNA by pullulan conjugation based on metal coordination, *J Control Release*, Vol. 83, 2002, pp. 287-302.

[67]   Jo, J., Ikai, T., Okazaki, A., Nagane, K., Yamamoto, M., Hirano, Y., Tabata, Y., Expression profile of plasmid DNA obtained using spermine derivatives of pullulan with different molecular weights, *J. Biomater. Sci. Polym Ed*, Vol. 18, 2007b, pp. 883-99.

[68]   Okazaki, A., Jo, J., Tabata, Y., A reverse transfection technology to genetically engineer adult stem cells, *Tissue Eng*, Vol. 13, 2007, pp. 245-251.

[69]   Jo, J., Okazaki, A., Nagane, K., Yamamoto, M., Tabata, Y., Preparation of cationized polysaccharides as gene transfection carrier for bone marrow-derived mesenchymal stem cells, *J Biomater Sci Polym Ed*, Vol. 21, 2010, pp. 185-204.

[70]   Gilchrist, R.K., Medal, R., Shorey, W.D., Hanselman, R.C., Parrot, J.C., Taylor, C.B., Selective inductive heating of lymph nodes, *Ann Surg*, Vol. 146, 1957, pp. 596-606.

[71]   Bulte, J.W. and Kraitchman, D.L., Monitoring cell therapy using iron oxide MR contrast agents. *Curr Pharm Biothecnol.*, Vol. 6, 2004, pp. 567-584.

[72]   Rodríguez-Luccioni, H.L., Latorre-Esteves, M., Méndez-Vega, J., Soto, O., Rodríguez, A.R., Rinaldi, C., Torres-Lugo, M., Enhanced reduction in cell viability by hyperthermia induced by SPIOs, *Int J Nanomedicine*, Vol. 6, 2011, pp. 373-380.

[73]   Ruan, J., Shen, J., Wang, Z., Ji, J., Song, H., Wang, K., Liu, B., Li, J., Cui, D., Efficient preparation and labeling of human induced pluripotent stem cells by nanotechnology, *Int J Nanomedicine*, Vol. 6, 2011, pp. 425-435.

[74]   Gordon, R., Hogan, C.E., Neal, M.L., Anatharam, V., Kanthsamy, A.G., Kanthasamy, A., A simple magnetic separation method for high-yield isolation of pure primary microglia, *J Neurosci methods*, 2011, 194, 287-296

[75]   Jain, T.K., Richey, J., Strand, M., Leslie-Pelecky, D.L., Flask, C.A., Labhasetwas, V., SPIOs with dual functional properties; drug delivery and magnetic resonance imaging. *Biomaterials*, Vol. 29, 2008, pp. 4012-4021.

[76]   Yallapu, M.M., Foy, S.P., Jain, T.K, Labhasetwar, V., PEG-functionalized SPIOs for drug delivery and magnetic resonance imaging applications, *Pharm Res,* Vol. 27, 2010, pp. 2283-2295.

[77]   Scherer, F., Anton, M., Schillinger, U., Henkel, J., Bergeman, C., Kruger, A., Gänsbacher, B., Plank, C., Magnetofection: enhancing and targeting

gene delivery by magnetic force in vitro and in vivo. *Gene Therapy*, Vol. 9, 2002, pp. 102-109.

[78]   Boyer, C., Whittaker, M.R., Bulmus, V., Liu, J., Davis, T.P., The design and utility of polymer-stabilized iron oxide nanoparticles for nanomedicine applications, *NPG Asia Mater*, Vol. 2, 2010, pp. 23-30.

[79]   Naqvi, S., Sanim, M., Abdin, M., Ahmed, F.J., Maitra, A., Prashant, C., Dinda, A.K., Concentration-dependent toxicity of iron oxide nanoparticles mediated by increased oxidative stress, Int. J. Nanomedicine, Vol.16, 2010, pp. 983-989.

[80]   Chertok, B., Moffat, B.A., David, A.E., Yu, F., Bregemann, C., Ross, B.D., Yang, V.C., Iron oxide nanoparticles as a drug delivery vehicle for MRI monitored magnetic targeting of brain tumours, *Biomaterials*, Vol. 29, 2008, pp. 487-96.

[81]   Kumar, A., Jena, P., Behera, S., Lockey, R., Mohapatra, S., Mohapatra, S., Multifunctional SPIOs for targeted delivery, *Nanomedicine*, Vol. 6, 2009, pp. 64-69.

[82]   Denizot, B., Tanguy, G., Hindre, F., Rump, E., Jeune, J., Jallet, P., Phosphorylcholine coating of iron oxide nanoparticles, *J Coll Interf Sci*, Vol. 209, 1999, pp. 66-71.

[83]   Ku, S., Yan, F., Wang, Y., Sun, Y., Yang, N., Ye, L., The blood-brain barrier penetration and distribution of PEGylated fluorescein-doped magnetic silica nanoparticles in rat brain, *Biochem Biophys Res Commun.*, Vol. 394, 2010, pp. 871-876.

[84]   Dias, A.M.G.C., Hussain, A., Marcos, A.S., Roque, A.C.A., A biotechnological perspective on the application of iron oxide magnetic colloids modified with polysaccharides, *Biotech adv*, Vol. 29, 2011, pp. 142-155.

[85]   Saiyed, Z.M., Gandhi, N.H., Nair, M.P.N., Magnetic nanoformulation of azidothymidine 5'-triphosphate for targeted delivery across the blood-brain barrier, *Int J Nanomedicine*, Vol. 5, 2010, pp. 157-66.

[86]   Vernon, H., Clarck, K., Bressler, J.P., In vitro models to study the blood brain barrier, *In vitro neurotoxicology: Methods and protocols*, 2010, Chapter 10, Vol. 758, pp. 153-168.

[87]   Deli, M. A., Ábrahám, C.S., Kataoka, Y., Niwa, M., Permeability studies on in vitro blood-brain barrier models: Physiology, pathology, and pharmacology, *Cell Mol Neurobiol*, Vol. 25, 2005, pp. 59-127.

[88]   Calibria, A.R., Weidenfeller, C., Jones, A.R., de Vries, H.E., Shusta, E.V., Puromycin-purified rat brain microvascular endothelial cell cultures exhibit

improved barrier properties in response to glucocorticoid induction, *J Neurochem*, Vol. 97, 2006, pp. 922-933.

[89] Lamszus, K., Schmidt, N.O., Ergün, S., Westphal, M., Isolation and Culture of Human Neuromicrovascular BCECs for the Study of Angiogenesis In Vitro, *J Neurosci Res*, Vol, 55, 1999, pp. 370-381.

[90] Bowman, P.D., Ennis, S.R., Rarey, K.Y., Betz, A.L., Goldstein, G.W., Brain microvessel BCECs in tissue culture: a model for study of blood-brain barrier properties, *Ann neurol*, Vol. 14, 1983, pp. 396-402.

[91] Roux, F., Durieu-Trautmann, O., Chaverot, N., Claire, M., Mailly, P., Bourre, J.M., Strosberg, A.D., Courad, P.O., Regulation of gamma-glytamyl transpeptidase and alkaline phosphatase activities in immortalized rat brain microvessel BCECs, *J Cell Physiol*, Vol. 159, 1994, pp. 101-113.

[92] Weksler, B.B., Subileau, E.A., Perriére, N., Charneau, P., Holloway, K., Leveque, M., Tricoire-Leignel, H., Nicotra, A., Bourdoulous, S., Turowski, P., Male, D.K., Roux, F., Greenwood, J., Romero, I.A., Couraud, P.O., Blood-brain barrier-specific properties of a human adult brain endothelial cell line, *FASEB J*, Vol. 19, 2005, pp. 1872-1874.

[93] Boado, R.J. and Pardridge, W.M., Measurement of blood- brain barrier GLUT1 glucose transporter and actin mRNA by a quantitive polymerase chain reaction assay, *J. Neurochem*, Vol. 62, 1994, pp. 2085-90.

[94] Rubin, L.L., Hall, D.E., Porter, S., Barbu, K., Cannon, C., Horner, H.C., Janatpour, M., Liaw, C.W., Manning, K., Morales, J., Tanner, L., Tomaselli, K.J., Bardet, F., A cell culture model of the blood brain barrier, *J Cell Biol*, Vol. 115, 1991, pp. 1725-1735.

[95] Nakagawa, S., Deli, M.A., Kawaguchi, H., Shimizudan, T., Shimono, T., Kittel, A., Tanaka, K., Niwa, M., A new blood-brain barrier model using primary rat brain endothelial cell, pericytes and astrocytes, *Neurochem Int*, Vol. 54, 2009, pp. 253-263.

[96] Hoeheisel, D., Nitz, T., Franke, H., Wegener, J., Hakvoort, A., Tilling, T., Galla, H.J., Hydrocortisone reinforces the blood-brain barrier proporties in a serum free cell culture system, *Biochem Biophys Res Commun*, Vol. 244, 1998, pp. 312-316.

[97] Dehouck, M.P., Méresse, S., Delorme, P., Fruchart, J.C., Cecchelli, R., An easier, reproducible, and mass-production method to study the blood-brain barrier in vitro, *J Neurochem*, Vol. 54, 1990, pp. 1798-1801.

[98] Gaillard, P.J., Voorwinden, L.H., Nielsen, J.L., Ivanov, A., Atsumi, R., Engman, H., Ringbom, C., de Boer, A.G., Breimer, D.D., Establisment and functional characterization of an in vitro model of the blood-brain barrier, comprising a co-culture of brain capillary BCECs and astrocytes, *Eur J Pharm Sci*, Vol. 12, 2001, pp. 215-222.

[99]   Tabell, J.M., Shear stress and the endothelial transport barrier, *Cardiovasc Res*, Vol. 87, 2010, pp. 320-330.

[100]   Cucullo, L., Hossain, M., Puvenna, V., Marchi, N., Janigro, D., The role of shear stress in blood-brain barrier endothelial physiology, *BMC Neurosci*, 2011, 12:40.

[101]   Stanness, K.A., Neumaier, J.F., Sexton, T.J., Grant, G.A., Emmi, A., Maris, D.O., Janigro, D., A new model of the blood-brain barrier: co-culture of neuronal, endothelial and glial cells under dynamic conditions, *Neuroreport*, Vol. 10, 1999, pp. 3725-3731.

[102]   Cucullo, L., McAllister, M.S., Knight, K., Krizanac-Bengez, L., Maroni, M., Mayberg, M.R., Staness, K.A., Janigro, D., A new dynamic in vitro model for the multidimensional study of astrocyte-endothelial cell interactions at the blood-brain barrier, *Brain Res*, Vol. 951, 2002, pp. 243-254.

[103]   Parkinson, F.E., Friesen, J., Krizanac-Bengez, L., Janigro, D., Use of a three-dimensional in vitro model of the rat blood-brain barrier to assay nucleoside efflux from brain, *Brain Res*, Vol. 980, 2003, pp. 233-241.

[104]   Krizanac-Bengez, L., Kapural, M., Parkinson, F., Cucullo, L., Hossain, M., Mayberg, M.R., Janigro, D., Effects of transient loss of shear on blood-brain barrier endothelium: role of nitric oxide and IL-6, *Brain Res*, Vol. 977, 2003, pp. 239-246.

[105]   Cucullo, L., Aumayr, B., Rapp, E., Janigro, D., Drug delivery and in vitro models of the blood-brain barrier, *Curr Opin Drug Disc Devel*, Vol. 8, 2005, pp. 89-99.

[106]   Santaguida, S., Janigro, D., Hossain, M., Oby, E., Rapp, E., Cucullo, L., Side by side comparison between dynamic versus static models of blood-brain barrier in vitro: permeability study, *Brain Res*, Vol. 1109, 2006, pp. 1-13.

[107]   Cucullo, L., Hossain, M., Rapp, E., Manders, T., Marchi, N., Janigro, D., Development of a humanized in vitro blood-brain barrier model to screen for brain penetration of antiepileptic drugs, *Epilepsia*, Vol. 48, 2007, pp. 505-516.

[108]   Cucullo, L., Couraud, P.O., Weksler, B., Romero, I.A., Hossain, M., Rapp, E., Janigro, D., Immortalized human brain BCECs and flow-based vascular modeling: a marriage of convenience for rational neurovascular studies, *J Cereb Blood Flow Metab*, Vol. 28, 2008, pp. 312-328.

[109]   Hossain, M., Sathe, T., Fazio, V., Mazzone, P., Weksler, B., Janigro, D., Rapp, E., Cucullo, L., Tobacco smoke: A critical etiological factor for vascular impairment at the blood-brain barrier, *Brain Res*, Vol. 1287, 2009, pp. 192-205.

[110] Jiang, C., Kouyabu, N., Yonemitsu, Y., Shimazoe, T., Watanabe, S., Naito, M., Tsuruo, T., Ohtani, H., Sawada, Y., In vivo delivery of glialcell derived neurothrophic factor across the blood-brain barrier by gene transfer into brain capillary BCECs, *Hum Gen Ther*, Vol. 14, 2003, pp. 1181-1191.

[111] Schabitz, W., Sommer, C., Zoder, W., Kiessling, M., Schwaninger, M., Schwab, S., Intravenous brain-derived neurotrophic factor reduces infarct size and counterregulates bax and bcl-2 expression after temporary focal cerebral ischemia, *Stroke*, Vol. 31, 1997, pp. 2212-2217.

[112] Makar, T.P., Bever, C.T., Singh, I.S., Royal, W., Sahu, S.N., Sura, T.P., Sultana, S., Sura, K.T., Patel, N., Dhib-Jalbut, S., Trisler, D., Brain-derived neurotrophic factor gene delivery in an animal model of multiple sclerosis using bone marrow stem cells as vehicle, *J Neuroimmunol*, Vol. 210, 2009, pp. 40-51.

[113] Johnston, P., Nam, M., Hossain, R., Indurti, R., Mankowski, J.L., Wilson, M.A., Laterra, J., Delivery of human fibroblast growth factor-1 gene to brain by modified rat brain endothelial cells, *J. Neurochem*, Vol. 67, 1996, pp. 1643-1652.

[114] Isgaard, J., Aberg, D., Nilsson, M., Protective and regenerative effects of the GH/IGF-1 axis on the brain, *Minerva Endocrinol*, Vol. 32, 2007, pp. 103-113.

[115] Wang, H., Chen, J., Sun, Y., Deng, J., Li, C.; Zhang, X., Zhuo, R., Construction of cell penetrating peptide vectors with N-terminal stearylated nuclear localization signal for targeted delivery of DNA into the cell nuclei, *J Control Release*, Vol. 155, 2011, pp. 26-33.

[116] Chowdhury, E.H., Nuclear targeting of viral and non-viral DNA, *Expert Opin Drug Deliv*, Vol. 6, 2009, pp. 697-703.

[117] Thakor, D.K., Teng, Y.D., Obata, H., Nagane, K., Saito, S., Tabata, Y., Nontoxic genetic engineering of mesenchymal stem cells using serum-compatible pullulan-spermine/DNA anioplexes, *Tissue Eng Part C Methods*, Vol. 17, 2011, pp. 131-144.

[118] Yang, H., Nanoparticle-mediated brain-specific drug delivery, imaging, and diagnosis, *Pharm Res*, Vol. 27, 2010, pp. 1759-71.

[119] Gaillard, P.J. and de Boer, A.G., Relationship between permeability of the blood-brain barrier and in vitro permeability coefficient of a drug, *Eur J Pharm Sci*, Vol. 12, 2000, pp. 95-102

[120] Helms, H.C., Waagepetersen, H.S., Nielsen, C.U., Brodin, B., Paracellular tightness and claudin-5 expression is increased in the BCEC/astrocyte blood-brain barrier model by increasing media buffer capacity during growth, *AAPS J*, Vol. 12, 2010, pp. 759-770.

[121] Nitz, T., Eisenblätter, T., Psathaki, K., Galla, H.J., Serum-derived factors weaken the barrier properties of cultured porcine brain capillary endothelial cells in vitro, *Brain Res*, Vol. 981, 2003, pp. 30-40.

[122] Nakagawa, S., Deli, M.A., Nakao, S., Honda, M., Hayashi, K., Nakaoke, R., Kataoka, Y., Niwa, M., Pericytes from brain microvessels strenghten the barrier integrity in primary cultures of rat brain BCECs, *Cell Mol Neurobiol*, Vol. 27, 2007, pp. 687-694.

[123] Skjørringe, T., Gjetting, T., Jensen, T.G., A modified protocol for efficient DNA encapsulation into pegylated immunoliposomes (PILs), *J Control Release*, Vol. 139, 2009, pp. 140-145.

[124] Young, V.W., Prospects of repair in multiple sclerosis, *J Neurol Sci*, Vol. 277, 2009, pp. 16-18.

[125] Zhu, J.M., Zhao, Y.Y., Chen, S.D., Zhang, W.H., Lou, L., Jin, X., Functional recovery after transplantation of neural stem cells modified by brain-derived neurotrophic factor in rats with cerebral ischaemia, *J Int Med Res*, Vol. 39, 2011, pp. 488-498.

[126] Yu, H. and Chen, Z., The role of BDNF in depression on the basis of its location in the neural circuitry, *Acta Pharmacol Sin*, Vol. 32, 2011, pp. 3-11.

[127] Sun, M., Kong, L., Wang, X., Lu, X.G., Gao, Q., Geller, A.L., Comparison of the capability of GDNF and BDNF or both to protect nigrostriatal neurons in a rat model of Parkinson's disease, *Brain Res*, Vol. 1052, 2005, pp. 119-129.

[128] Li, L., Li, J., Wu, Q., Li, J., Feng, Z., Liu, S., Wang, T., Transplantation of NGF-gene-modified bone marrow stromal cells into a rat model of Alzheimer' disease, *J Mol Neurosci*, Vol. 34, 2008, pp. 157–163.

[129] Zhang, T., Qu, H., Li, X., Zhao, B., Zhou, J., Li, Q., Sum, M., Transmembrane delivery and biological effect of human growth hormone via a phage displayed peptide in vivo and in vitro, *J Pharm Sci*, Vol. 99, 2010, pp. 4880-4891.

[130] Song, B., Vinters, H.V., Wu, D., Pardridge, W.M., Enhanced neoroprotective effects of basic fibroblast growth factor in regional brain ischemia after conjugation to a blood-brain barrier vector, *J Pharmcol*, Vol. 301, 2002, pp. 605-610.

[131] Ma, Y.P., Ma, M.M., Cheng, S.M., Ma, H.H., Yi, X.M., Xu, G.L., Liu, X.F., Intranasal bFGF-induced progenitor cell proliferation and neuroprotection after transient focal cerebral ischemia, *Neurosci let*, Vol. 437, 2008, pp. 93-97.

[132] Boado, R.J., Hui, E.K., Lu, J.Z., Pardridge, W.M., Drug targeting of erythropoietin across the primate blood-brain barrier with an IgG molecular trojan horse, *J Pharmacol Exp Ther*, Vol. 333, 2010, pp. 961-969.

[133] Xue, Y.Q., Zhao, L.R., Guo, W.P., Duan, W.M., Intrastriatal administration of erythropoietin protects dopaminergic neurons and improves neurobehavioral outcome in a rat model of Parkinson's disease, *Neurosci*, Vol. 146, 2007, pp. 1245-1258.

[134] Iwai, M., Stetler, R.A., Xing, J., Hu, X., Gao, Y., Zhang, W., Chen, J., Cao, G., Enhanced oligodendrogenesis and recovery of neurological function by erythropoietin following hypoxic/ischemic brain injury, Stroke, Vol. 41, 2010, pp. 1032-1037.

[135] Lundberg, C., Bjorklund, T., Carlsson,T., Jakobsson, J., Hantraye, P., Deglon, N., Kirik, D., Applications of lentiviral vectors for biology and gene therapy of neurological disorders, *Curr Gene Ther*, Vol. 8, 2008, pp. 461-473.

[136] Gray, S.J., Woodard, K.T., Samulski, R.J., Viral vectors and delivery strategies for CNS gene therapy, *Ther Deliv*, Vol. 1, 2010, pp. 517-534.

[137] Zhang, Y. and Yu, L.C., Microinjection as a tool of mechanical delivery, *Curr Opin Biotech*, Vol. 19, 2008, pp. 506-510.

[138] Benediktsson, A.M., Schachtele, S.J., Green, S.H., Dailey, M.E., Ballistic labeling and dynamic imaging of astrocytes in organotypic hippocampal slice culture, *J Neurosci Methods*, Vol. 141, 2005, pp. 41-53.

[139] De Vry, J., Martinez-Martinez, P., Losen, M., Temel, Y., Steckler, T., Steinbusch, H.W.M., De Baets, M.H., Prickaerts, J., In vivo electroporation of the central nervous system: a non-viral approach for targeted gene delivery, *Prog Neurobiol*, Vol. 92, 2010, pp. 227-244.

[140] Tros de Ilarduya, C., Sun, Y., Düzgünes, N., Gene delivery by lipoplexes and polyplexes, *Eur J Pharm Sci*, Vol. 40, 2010, pp. 159-170.

[141] Cavaletti, G., Cassetti, A., Canta, A., Galbiati, S., Gilardini, A., Oggioni, N., Rodriguez-Menendez, V., Fasano, A., Liuzzi, G.M., Fattler, U., Ries, S., Nieland, J., Riccio, P., Haas, H., Cationic liposomes target sites of acute neuroinflammation in experimental autoimmune encephalomyelitis, *Mol Pharm*, Vol. 6, 2009, pp. 1363-70.

[142] Lehtinen, J., Hyvönen, Z., Subrizi, A., Burnjes, H., Urtti, A., Glycosaminoglycan-resistant and ph-sensitive lipid coated DNA complexes produced by detergent removal method, *J Control Release*, Vol. 131, 2008, pp. 145-149.

[143] Son, S., Hwang do, W., Singha, K., Jeong, J.H., Park, T.G., Lee, D.S., Kim, W.J., RVG peptide tethered bioreducible polyethylenimine for gene delivery to the brain, *J Control Release*, Vol. 155, 2011, pp. 18-25.

[144] Svenson, S., Dendrimers as versatile platform in drug delivery applications, *Eur J Pharm BioPharm*, Vol. 71, 2009, pp. 445-462.

[145] Shao, K., Huang, R., Li, J., Han, L., Ye, L., Lou, J., Jiang, C., Angiopep-2 modified PE-PEG based polymeric micelles for amphotericin B delivery targeted to the brain, *J Control Release*, Vol. 147, 2010, pp. 118-126.

Appendix I

3330   *Current Medicinal Chemistry*, 2011, *18*, 3330-3334

Nanoparticle-Derived Non-Viral Genetic Transfection at the Blood-Brain Barrier to Enable Neuronal Growth Factor Delivery by Secretion from Brain Endothelium

L.B. Thomsen, A.B. Larsen, J. Lichota and T. Moos*

Section of Neurobiology, Biomedicine, Aalborg University, Denmark

Abstract: Brain capillary endothelial cells form the blood-brain barrier (BBB) that denotes a major restraint for drug entry to the brain. The identification of many new targets to treat diseases in the brain demands novel thinking in drug design as new therapeutics could often be proteins and molecules of genetic origins like siRNA, mRNA and cDNA. Such molecules are otherwise prevented from entry into the brain unless encapsulated in drug carriers. The desirable entry of such large, hydrophilic molecules should be made by formulation of particular drug carriers that will enable their transport into the brain endothelium, or even through the endothelium and into the brain. This manuscript reviews the potential of different drug-carriers for therapy to the brain with respect to their targetability, biocompatibility, toxicity and biodegradability.

Keywords: Blood-brain barrier, Drug delivery, Lipoplexes, Magnetic nanoparticles, Nanoparticle, Polyplexes, Pullulan, Transferrin receptor.

INTRODUCTION

Drug delivery to the brain is restricted by the barriers of the brain, i.e. the blood-brain barrier (BBB) and the blood-cerebrospinal fluid (CSF) barrier (Fig. 1). The relative surface area of the capillaries in the brain covered by the blood- CSF barrier is very small compared to that covered by the BBB, which makes the BBB the main route of interest for systemic drug delivery to the brain [1]. The BBB maintains the homeostasis of solutes in the central nervous system (CNS) and prevents potentially harmful substances from entering the brain from the systemic circulation. Only a few percentages of drugs intended for treatment of diseases in the CNS can pass the BBB to enter the brain parenchyma. The molecules most likely to penetrate the BBB are small-sized (less than 70 Dalton), and either highly lipophilic or with good affinity for endogenous transporter molecules expressed by the BBB [2].

Several existing and approaching drug candidates are composed of amino acids (proteins) or nucleic acids (cDNA, miRNA, siRNA), which are all hampered by a complete inability to pass the brain barriers due to the size and hydrophilic nature [2]. Nonetheless, this unavailability to enter the brain has sprouted countless approaches to design molecular-carriers that could enable efficient concentrations of therapeutics inside the brain. The yield of sufficient amounts of protein could theoretically be obtained by transport of nanoparticles containing proteins across the BBB or by transfection of the brain by particles containing cDNA encoding proteins that would be expressed once internalized by cells of the brain.

The evidence that nanoparticles containing proteins can enter the brain and lead to significant concentrations inside the brain is controversial. The foremost evidence presented rapports on transport of carriers of genetic material that transfect the brain to enable cells of the brain to become a local protein source. Concerning the carriers of genetic material, viral vectors have shown to be efficient carriers although they are hampered by the risks of causing immunogenicity and to integrate in places of the human genome that potentially could cause cancer [3].

As an alternative to viral gene therapy, non-viral vectors are considered as biocompatible and often non-immunogenic. They also exert low toxicity, but they have lower transfection efficiency than their viral counterparts. Preferably, a desirable drug delivery system should be a non-viral vector with transfection efficiency equally high to the viral vector. It should also be biodegradable, non-immunogenic and designed to allow for a selected therapy to cells of interest expressing a particular targeting molecule.

Several approaches have been made to evolve this non-viral vector delivery transport system with high efficiency. In 1974, Gregoriadis *et al.* suggested that liposomes could be used as carriers for drug delivery [4]. This approach was later modified and expanded to the use of immunoliposomes and so-called lipoplexes and probably remains the most successful strategy within the field of non-viral gene therapy. Other approaches have been applied to prepare carrier complexes from electrostatic binding properties of cationic polymers that will attach to the highly anionic binding sites of nucleic acids [5]. More recently, carriers based on magnetic nanoparticles that can be very accurately delivered by the use of an external magnetic field are being developed [6,7]. This review covers the BBB and the obstacles involved in drug delivery to the brain. The most promising non-viral drug delivery transport systems to the BBB, i.e. nanoparticle technology based on liposomes, cationic polymers and magnetic nanoparticle, are outlined and recent progresses within these systems are addressed.

THE BLOOD-BRAIN BARRIER

The BBB formed by the endothelial cells of the brain capillaries constitutes a physical, chemical and immunological barrier between the blood and the brain. The endothelial cells make intimate contacts with astrocytes, pericytes and perivascular macrophages that all constitute a significant role in maintaining structural and permeability properties of the BBB. Together with the perivascular astrocytes the endothelial cells synthesize a basal lamina, which mainly consists of laminin, type-IV collagen, integrins and fibronectin [2,8]. Althougha complete barrier, it is believed that the basal lamina limits the passage of macromolecules and particles if they get from the endothelial cells into the brain [2]. Astrocytic endfeet make contact with the basal lamina and envelope almost the entire abluminal surface of the microvessels. The astrocytes are important for induction and regulation of the BBB properties of the endothelial cells. The pericytes are embedded in the basal lamina and cover up to 30 % of the microvessel surface. They presumably have a role in regulating the paracellular permeability of the BBB by regulating the functioning of tight junctions between the endothelial cells [9].

The brain capillary endothelial cells (BCECs) are of the non-fenestrated type, rich in mitochondria but low in pinocytotic transport activity. The tight junctions between the endothelial cells limit the paracellular trafficking of ions and macromolecules [8]. Transcellular vesicular transport is also considered limited as influx and efflux transporters regulate the main entry and exit of molecules considered transported in non-vesicular conditions.

Different transcellular transport mechanisms allow some molecules to pass the BBB. Facilitated by a concentration gradient, lipophilic and non-polar molecules diffuse passively across the brain capillaries. Solute carrier transporters allow for transport across the

*Address correspondence to this author at the Section of Neurobiology, Biomedicine, Department of Health Science and Technology, Fr. Bajers Vej 3B, 1.216, Aalborg University, DK-9220 Aalborg East, Denmark; Tel: + 45-99442420; Fax: (45) 96357816; E-mail: tmoos@hst.aau.dk

0929-8673/11 $58.00+.00

Drug Delivery to the Brain   Current Medicinal Chemistry, 2011 Vol. 18, No. 22   3331

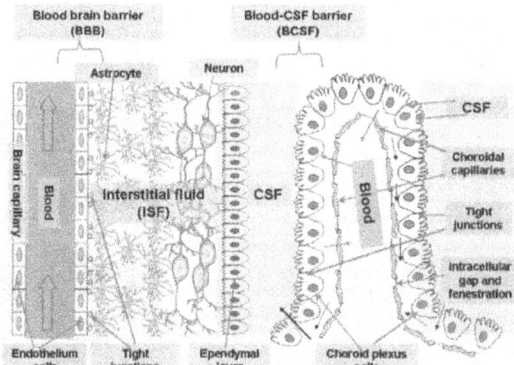

Fig. (1). Overview of the two main barriers in the CNS. blood-brain barrier and blood cerebrospinal fluid barrier (BCSF). ISF: Interstitial Fluid. CSF: Cerebrospinal fluid. [39, Open Access Journal. Publisher BioMed Central].

BBB of nutrients and other essential molecules like glucose, amino acids and nucleosides. Macromolecules, like insulin, transferrin, low-density lipoprotein and albumin, are taken up by brain endothelial cells from the plasma by endocytosis either via receptor mediated endocytosis or adsorptive mediated endocytosis, but whether they undergo further transport through the endothelial cells into the BBB has not been proved [2]. Active efflux transporters like P-glycoprotein (P-gp) and multidrug resistance associated protein 1 (MRP1) are members of the ATP-binding cassette (ABC) transporters. They are present both in the luminal and abluminal membrane of the endothelial cells and extrude substances that are not recognized as being essential for the brain irrespective of size [8]. Not only do the endothelial cells have a high metabolic activity indicated by their richness in mitochondria, but they also have high enzymatic activity that degrades peptides and other compounds of the plasma [10].

TARGETED THERAPY AT THE BLOOD-BRAIN BARRIER

As the passage of drugs across the BBB is limited to small hydrophobic molecules or molecules with structural similarity of nutrients and high affinity for endogenous transporters, the transport of large hydrophilic molecules with therapeutic potential (see above) needs the approach of a drug carrier. Preferably, the carrier should be conjugated with a targeting molecule to facilitate the uptake by the endothelial cells from the plasma. The selection of molecules for targeted therapies at the BBB was recently reviewed [c.f. 2]. Targeting endogenous molecules expressed on the luminal side of the BCECs can be achieved with peptides or derivates, peptidomimetics. Alternatively, anti-receptor antibodies against the transferrin receptor or the insulin receptor can be used to target the brain capillaries. The anti-transferrin receptor antibody is of particular interest as the receptor is exclusively expressed in BCECs and not in the endothelium of any other organ in the body [11].

Evidently large hydrophilic molecules, such as albumin and IgG, are capable of passing through the blood-CSF barrier through choroid plexus epithelial cells by means of transcytosis [2]. From a drug-delivery perspective, this route however is considered less relevant, as the relative surface area of the blood-CSF barrier com-

pared to the brain capillaries is very low. Moreover, blood-CSF transport leads to the appearance of molecules in the ventricular system and not in the brain interstitium from where uptake by neurons would be much higher (Fig. 1). Possibly direct injection into the ventricular system of drug-carriers without targeting molecule can be used for drug transport into the brain [2].

TOOLS FOR NON-VIRAL DRUG DELIVERY: LIPID-BASED-, CHARGED - AND MAGNETIC NANOPARTICLES

Liposomes, Immunoliposomes and Lipoplexes

Liposomes are biocompatible, virtually non-toxic and biodegradable. They consist of a bilayer of phospholipids or sphingolipids that form uni- or multilaminar spheres [10]. The liposomes can be used for encapsulating both lipophilic and hydrophilic substances, which can be carried either within the liposomal spheres or on their surface. The durability of liposomes in blood plasma is limited because of their rapid clearance by phagocytotic cells. This difficulty can be circumvented by coating the liposomes with polyethyleneglycol (PEG), polypropylene oxide or another surface modifying substrate leading to steric stabilization of the liposomes [12].

Liposomes can be delivered to cells in a targetable manner. The coating molecule PEG can bridge the binding between the liposome and a targetable antibody, hence turning the liposome into a targetable immunoliposomes [10,13]. The presence of the targetable molecule will enable the immunoliposomes to enter the cell via targeted delivery trough receptor-mediated endocytosis [14]. For targeted delivery to BCECs, the most prevalently used antibody is raised against the transferrin receptor, which is abundant in the BCECs [1,2,13,15].

A novel approach has been to link the liposome with two different antibodies, the approach being that this will increase the attachment to both an extracellular epitope and a molecule of the intracellular compartment to improve the efficiency and delivery rate of the liposome and its content [16-22]. Dual ligand-coupled liposomes have also been used for both enhancing the delivery to the brain in nude mice implanted with human cancer cells by cou-

3332 *Current Medicinal Chemistry, 2011 Vol. 18, No. 22*   *Thomsen et al.*

pling liposomes to the antibody against the mouse transferrin receptor, thereby targeting the endothelial cells, and the human insulin receptor, thereby targeting the cancer cells [21]. Specificity of the transfection with immunoliposomes carrying plasmid DNA was improved by adding a brain specific promoter to the cDNA. The cDNA carried within the immunoliposome was selectively expressed by cells of the brain [23-25], e.g. inclusion of the promoter regulating the expression of glial fibrillary acidic protein (GFAP) lead to its specific expression in astrocytes of the CNS [25].

Lipoplexes are complexes of cationic lipids and DNA, which are formed by electrostatic interactions between cationic lipids and the anionic DNA [26]. The DNA inside the lipoplexes is arranged in clusters of parallel threads when absorbed on the surface of the lipids or trapped in the lipid bilayer [27]. The size of the lipoplexes seems to influence the transfection efficiency. A diameter ranging from 0.4-1.4 μm has shown to be the optimal size for complexes containing cholesterol. An obvious challenge in using the lipoplexes is the limitation in the capacity to bind DNA as the lipoplexes are rather small. Another problem is the virtual dose dependent toxicity of these cationic lipids that could limit their therapeutic use [27].

Cationic Polymers, Polyplexes, for Targeting Purposes

Complexes of cationic polymers and DNA are generally referred to as polyplexes. The polymers may consist of endogenously synthesized peptides, proteins, polyamines and/or polysaccharides [28,29,30]. They have been proven to be potent carriers of nucleic acids because of their high ability to bind and condense the anionic plasmid DNA electrostatically. Synthetic cationic polymers such as polyethylenimine (PEI), polysaccharides, and polypeptides are also being used for transfection purposes but a main concern on the use of synthetic polyplexes is their toxicity, e.g. PEI was shown to induce cell death in cultured cells [31]. A principle difference between polyplexes and liposomes is that polyplexes cannot directly release their DNA load into the cytoplasm as they enter the cells by endocytosis. To release from endosomes, polyplexes like PEI can cause the endosome to swell, which will lead to its degradation. But for obvious reasons, this degradation of the endosomal compartment is also responsible for toxicity of the cell. The endosomal rupture may further induce PEI to exert its cytotoxicity to cells by two different mechanisms that can lead to rapid cell death due to necrosis, either due to loss of the cell membrane integrity or by a slower damaging process leading to apoptosis due to loss of the mitochondrial membrane potential [32].

The obstacle of the toxicity of synthetic polyplexes can be overcome by the use of natural cationic polymers, which are non-toxic or a combination of synthetic and natural polymers. Poly(lactic acid) (PLA) is biodegradable, low in toxicity and has good mechanical properties that, together with its Tween-80 (polysorbate 80) coat, eases its cellular permeability [33]. The polysorbate coat interacts with apolipoproteins B and E *in vivo*, and hence interacts indirectly with the receptors of these lipoproteins [c.f. 2]. Albumin-based apolipoproteins E nanoparticles can enter the brain *via* transport across the BBB, thus making it a good promising vehicle for drug delivery to the brain, but as these nanoparticles also interfere with capillary endothelial cells elsewhere in the body, it remains to be improved regarding its specificity for delivery to the brain [34].

A novel drug carrier complex pullulan-spermine complexed with plasmid DNA shows promising potential for usage for transfection of brain endothelial cells [5]. Pullulan is a water soluble polysaccharide, and spermine is a naturally occurring polyamine present in all eukaryotic cells and is involved in basic cellular metabolism [35,36]. Pullulan-spermine is known to undergo cellular endocytosis *via* clathrin or raft/caveolae dependent endocytosis. Using plasmids containing cDNA coding for HcRed fluorescent protein as a

reporter gene [cf. 5], BCECs were shown *in vitro* to secrete human growth hormone 1 (hGH1) after transfection (Figs. 2,3). The pullulan-spermine carrier complex is also suitable for conjugation with an antibody raised against the transferrin receptor making it suitable for targeting purposes [5].

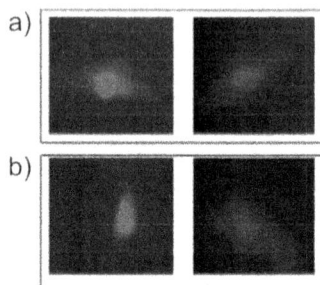

Fig. (2). Pullulan transfected brain capillary endothelial cells expressing Hc-Red. RBE4s (a) and HBMECs (b) were transfected with pullulan-spermine conjugated with pHc-Red1-C1, a red fluorescent reporter gene. Note that the cells express Hc-red in both the cell cytosol and nucleus (*). The cells were observed in a fluorescence microscope with 400x magnification [5].

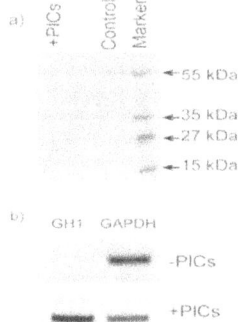

Fig. (3). Expression of human growth hormone in HBMECs. a) Detection of FLAG-tagged hGH1 on a PVDF membrane with TMB. First lane (+PICs) shows immunoprecipitated FLAG-tagged hGH1 proteins from pullulan transfected HBMECs. A band is seen just below 27 kDa, corresponding to the FLAG-tagged hGH1, which has a size of 23 kDa. Second lane (-PICs) shows immunoprecipitate from non-transfected HBMECs and no band is seen. Third (control) immunoprecipitated FLAG-BAP fusion protein, which normally migrates as a 45-55 kDa band. The fourth lane (marker) is a prestained protein ladder. b) RT-PCR analysis of human hGH1 expression after pullulan transfection of HBMECs. hGH1 transcripts were clearly present in transfected HBMECs (+PICs), whereas in non-transfected HBMECs (-PICs) a vague amount of hGH1 transcripts was seen [5].

Magnetic Nanoparticles

In medical sciences, magnetic nanoparticles are being used for purposes like magnetic resonance imaging (MRI), hyperthermia for tumor therapy, drug delivery and targeted therapy. They have been successfully used for delivery of anticancer drugs in treatment of brain carcinomas where the BBB is compromised [37].

Magnetic nanoparticles normally consist of a core of nano-sized iron-oxide particles such as magnetite (Fe_3O_4) and maghemite(γ-Fe_2O_3). This magnetic core can be coated with a fluorescent dye and these components are either covered by a biocompatible polymeric shell, like dextran, polysorbate or starch, or coated by phospholipids hence creating a magneto-polyplex or a magneto-liposome [6,37]. Like liposomes, magnetic nanoparticles must be modified with e.g. PEG or dextran as they otherwise are prone to rapid clearance from the systemic circulation due to uptake by macrophages [38]. Coating reduces toxicological effects by hindering leaching of non-biodegradable magnetic cores and oxidation. A surface coat of hydrophilic polymers also minimizes aggregation of magnetic nanoparticles which can otherwise lead to embolisms of capillaries [37]. Coated magnetic particles are furthermore suitable for targeting to various cellular surface proteins.

The magnetic nanoparticles may also be very precisely delivered to a target organ with the aid of a magnetic field [6,37]. This strategy of controlled drug delivery has e.g. been widely used for delivery of chemotherapeutics like doxorubicin and has now been made commercially available by e.g. Chemicell GMBH, Germany. The magnetic field is supplied by an external or an implanted magnet. When applied, the magnetic nanoparticles will be drawn towards the magnet and concentrate in the area where the magnet is located. The delivery will therefore be very local and the drug dosage can be minimized and side-effects can be reduced.

CONCLUSIONS

The optimal drug carrier can be characterized by means of its capability to be biocompatible, virtually non-toxic and biodegradable. It should also be targetable, able to bind a significant amount of cDNA, and able to transfect cells *in vivo*. We have found that pullulan-spermine complexed with plasmid DNA shows promising potential for usage for transfection of brain capillary endothelial cells and for using these cells as factories for protein secretion [5]. We are in the process of examining this complex for *in vivo* transfection in the brain. We have also found that magnetic particles may denote a novel delivery mechanism for neuronal gene therapy provided that the magnetic particles can be designed to enable transport from endothelium-to-brain [7].

ACKNOWLEDGEMENTS

The most recent results obtained and described by the authors were generated by generous grant support from the Danish Medical Research Council (grant no. 271-06-0211), the Spar Nord Fund, and the Obelske Family Fund.

ABBREVIATIONS

BBB	=	blood-brain barrier
CSF	=	Cerebrospinal fluid
CNS	=	central nervous system
BCECs	=	Brain Capillary Endothelial Cells
PEG	=	Polyethyleneglycol
PEI	=	Polyethylenimine

REFERENCES

[1] Pardridge, W.M. Brain drug targeting. *Cambridge University Press*, UK, 2001.

[2] Lichota, J.; Skjorringe, T.; Thomsen, L.B. Moos, T. Macromolecular drug transport into brain using targeted therapy *J. Neurochem*, 2010, 113, 1-13.

[3] Davis, M.E; Pun, S.H.; Bellocq, N.C.; Reineke, T.M.; Popielarski, S.R.; Mishra, S.; Heidel, J.D. Self-assembling nucleic acid delivery vehicles *via* linear, water-soluble, cyclodextrin-containing polymers, *Curr. Med. Chem.*, 2004, 11, 179-97.

[4] Gregoriadis, G.; Wills, E.J.; Swain, C.P.; Tavill, A.S. Drug-carrier potential of liposomes in cancer chemotherapy, *Lancet*, 1974, 1, 1313-6.

[5] Thomsen, L.B.; Lichota, J.; Kim, K.S.; Moos, T. Gene delivery by pullulan derivatives in brain capillary endothelial cells for protein secretion, *J. Control. Release*, 2011, 151, 45-50.

[6] Saiyed, Z.M.; Gandhi, N.H.; Nair, M.P.N. Magnetic nanoformulation of azidothymidine 5'-triphosphate for targeted delivery across the blood-brain barrier, *Int J Nanomedicine*, 2010, 5, 157-66.

[7] Thomsen, L.B.; Lichota, J.; Larsen T.E.; Linemann, T.; Mortensen, J.H. Nielsen, K.G.D.J.; Moos, T. Brain delivery systems *via* mechanisms independent of receptor-mediated endocytosis and adsorptive mediated endocytosis, *Curr Pharm Biotechnol*, 2011, In print.

[8] Abbott, N.J.; Patabendige, A.A.; Dolman, D.E.; Yusof, S.R.; Begley, D.J. Structure and function of the blood-brain barrier. *Neurobiol. Dis.* 2010, 37, 13-25.

[9] Persidsky, Y.; Ramirez, S.H.; Haorah, J.; Kanmogne, G.D. Blood-brain barrier: structural components and function under physiologic and pathologic conditions *J. Neuroimmune Pharmacol.*, 2006, 1, 223-36.

[10] Witt, K.A.; Gillespie, T.J.; Huber, J.D.; Egleton, R.D.; Davis, T.P. Peptide drug modifications to enhance bioavailability and blood-brain barrier permeability, *Peptides*, 2001, 22, 2329-43.

[11] Jefferies, W.A.; Brandon, M.R.; Hunt, S.V.; Williams, A.F.; Gatter, K.C.; Mason, D.Y. Transferrin receptor on endothelium of brain capillaries. *Nature*, 1984, 312, 162-63.

[12] Alam, M.L.; Beg, S.; Samad, A.; Baboota, S.; Kohli, K.; Ali, J.; Ahuja, A.; Akbar, M. Strategy for effective brain drug delivery, *Eur J Pharm Sci*, 2010, 40, 385-403.

[13] Yoshikawa, B.U.; Pardridge, W.M. Delivery of peptides and proteins through the blood-brain barrier, *Adv Drug Deliv Rev.*, 2001, 46, 247-79.

[14] Jølck, R.I.; Feldborg, L.N.; Andersen, S.; Moghimi, S.M.; Andresen, T.L. Engineering liposomes and nanoparticles for biological targeting, *Adv Biochem Engin Biotechnol*, 2010, epub ahead of print.

[15] Gosk, S.; Vermehren, C.; Storm, G.; Moos, T. Targeting anti-transferrin receptor antibody (OX26) and OX26-conjugated liposomes to brain capillary endothelial cells using in situ perfusion. *J. Cereb. Blood Flow Metab* 2004, 24, 1193-1204.

[16] Markoutsa, E.; Pampalakis, G.; Niarakis, A.; Romero, I.A.; Weksler, B.; Couraud, P.O.; Antimisiaris, S.G. Uptake and permeability studies of BBB-targeting immunoliposomes using the hCMEC/D3 cell line, *Eur J. Pharm Biopharm* 2010, Epub ahead of print.

[17] Ying, X.; Wen, H.; Lu, W.L.; Du, J.; Guo, J.; Tian, W.; Men, Y.; Zhang, Y.; Li, R.J.; Yang, T.Y.; Shang, D.W.; Lou, J.N.; Zhang, L.R.; Zhang, Q. Dual-targeting daunorubicin liposomes improve the therapeutic efficacy of brain glioma in animals, *J. Control. Release*, 2010, 141, 183-92.

[18] Safavy, A.; Raisch, K.P.; Matusiak, D.; Bhatnagar, S.; Helson, L. Single-drug multiligand conjugates: synthesis and preliminary cytotoxicity evaluation of a paclitaxel-dipeptide "scorpion" molecule, *Bioconjugate Chem.*, 2006, 17, 565-70.

[19] Andresen, T.L.; Jensen, S.S.; Jørgensen, K.; Advanced strategies in liposomal cancer therapy: problems and prospects of active and tumour specific drug release, *Prog. Lipid Res.*, 2005, 44, 68-97.

[20] Strukel, J.M.; Li, R.C.; Maynard, H.D.; Caplan, M. Two-step synthesis of multivalent cancer-targeting constructs, *Biomacromolecules*, 2010, 11, 160-67.

[21] Zhang, Y.; Zhu, C.; Pardridge, W.M. Antisense gene therapy of brain cancer with an artificial virus gene delivery system, *Mol. Ther.*, 2002, 6, 67-72.

[22] Tan, P.H.; Manunta, M.; Ardjomand, N.; Xue, S.A.; Larkin, D.F.; Haskard, D.O.; Taylor, K.M.; George, A.J. Antibody targeted gene transfer to endothelium, *J. Gene Med.*, 2003, 5, 311-23.

[23] Shi, N.; Zhang, Y.; Zhu, C. Brain-specific expression of an exogenous gene after i.v. administration. *Proc. Natl Accad Sci USA*, 2001, 98, 12754-59.

[24] Shu, C.; Zhang, Y.; Zhang, Y.F.; Yi, L.I.; Boado, R.J.; Pardridge, W.M. Organ-Specific expression of the LacZ gene controlled by the opsin promoter after intravenous gene administration in adult mice. *J Gene Med*, 2004, 6, 906-12.

[25] Zhao, H.; Li, G.L.; Wang, R.Z.; Li, S.F.; Wei, J.J.; Feng, M.; Zhao, Y.J.; Ma, W.B.; Yang, Y.; Li, Y.N.; Kong, Y.G. A comparative study of transfection efficiency between liposomes, immunoliposomes and brain-specific immunoliposomes, *J. Int. Med. Res.*, 2010, 38, 957-66.

[26] Elouahabi, A.; Ruysschaert, J.M. Formation and intracellular trafficking of liposomes and polyplexes, *Mol. Ther.*, 2005, 11, 336-47.

[27] Tarahovsky, Y.S. Cell transfection by DNA-lipid complexes – lipoplexes, *Biochemistry (Mosc.)*, 2009, 74, 1293-04.

3334   *Current Medicinal Chemistry, 2011 Vol. 18, No. 22*   *Thomsen et al.*

[28] Tiwari, S.B.; Amiji, M.M. A review of nanocarrier-based CNS delivery systems. *Cur. Drug. Deliv.*, **2006**, 3, 219-32.

[29] Teixido, M.; Giralt, E. The role of peptides in blood-brain barrier nanotechnology. *J. Pept. Sci.*, **2008**, 14, 163-73.

[30] Zhang, H.; Vinogradov, S.V. Short degradable polyamines for gene delivery and transfection of brain capillary endothelial cells, *J. Control. Release*, **2010**, 143, 359-66.

[31] Kim, I.B.; Choi, J.S.; Nam, K.; Lee, M.; Park, J.S.; Lee, J.K. Enhanced transfection of primary cortical cultures using arginine-grafted PAMAM dendrimer, PAMAM-Arg. *J. Control Release*, **2006**, 114, 110-17.

[32] Moghimi, S.M.; Symonds, P.; Murray, J.C.; Hunter, A.C.; Debska, G.; Szewczyk, A. A two-stage poly(ethylenimine)-mediated cytotoxicity: implications for gene transfer/therapy. *Mol. Ther.*, **2005**, 11, 990-95.

[33] Ren, T.; Xu, N.; Cao, C.; Yuan, W.; Yu, X.; Chen, J.; Ren, J. Preparation and Therapeutic Efficacy of Polysorbate-80-Coated Amphotericin B/PLA-b-PEG Nanoparticles. *J. Biomater.Sci. Polym. Ed.*, **2009**, 20, 1369-80.

[34] Zensi, A.; Begley, D.; Pontikis, C.; Legros, C.; Mihoreanu, L.; Wagner, S.; Büchel, C.; von Briesen, H.; Kreuter, J. Albumin nanoparticles targeted with Apo E enter the CNS by transcytosis and are delivered to neurones. *J. Control Release*, **2009**, 137, 78-86.

[35] Jo, J.; Ikai, T.; Okazaki, A.; Yamamoto, M.; Hirano, Y.; Tabata, Y. Expression profile of plasmid DNA by spermine derivatives of pullulan with different extents of spermine introduced. *J. Control. Release*, **2007**, 118, 389-98.

[36] Jo, J.; Ikai, T.; Okazaki, A.; Nagane, K.; Yamamoto, M.; Hirano, Y.; Tabata, Y. Expression profile of plasmid DNA obtained using spermine derivatives of pullulan with different molecular weights. *J. Biomater. Sci. Polym. Ed.*, **2007**, 18, 883-99.

[37] Chertok, B.; Moffat, B.A.; David, A.E.; Yu, F.; Bregemann, C.; Ross, B.D.; Yang, V.C. Iron oxide nanoparticles as a drug delivery vehicle for MRI monitored magnetic targeting of brain tumours. *Biomaterials*, **2008**, 29, 487-96.

[38] Ku, S.; Yan, F.; Wang, Y.; Sun, Y.; Yang, N.; Ye, L. The blood-brain barrier penetration and distribution of PEGylated fluorescein magnetic silica nanoparticles in rat brain. *Biochem. Biophys. Res. Commun.*, **2010**, 394, 871-6.

[39] Bhaskar, S.; Tian, F.; Stoeger, T.; Kreyling, W.; de la Fuente, J.M.; Grazu, V.; Borm, P.; Estrada, G.; Ntziachristos, V.; Razansky, D. Multifunctional Nanocarriers for diagnostics, drug delivery and targeted treatment across blood-brain barrier: perspectives on tracking and neuroimaging. *Part Fibre Toxicol*, **2010**; 7:3.

Received: March 09, 2011   Revised: May 30, 2011   Accepted: June 01, 2011